Our Science

Trinidad and Tobago

David Bainton

Jerome Ramdahin
Waterloo High School

Shameem Narine
La Romaine High School

CAMBRIDGE UNIVERSITY PRESS
Cambridge, New York, Melbourne, Madrid, Cape Town, Singapore,
São Paulo, Delhi, Dubai, Tokyo

Cambridge University Press
The Edinburgh Building, Cambridge CB2 8RU, UK

www.cambridge.org
Information on this title: www.cambridge.org/9780521607162

First published 2009

Printed in the United Kingdom by Printondemand

A catalogue record for this publication is available from the British Library

ISBN 978-0-521-60716-2 Paperback

Cover designer: Karen Ahlschläger
Typesetters: Charlene Bate and Nazley Samsodien
Illustrators: Robert Hichens, Saaid Rahbeeni, Geoff Walton and James Whitelaw

Contents

What is an ecosystem?

Keywords

abiotic, aquarium, biotic, ecology, ecosystem, evolution, interaction, terrarium

Did you know?

Gaia theory was developed by the scientist James Lovelock to help understand global climate change. Lovelock says that we should think of both living and non-living parts of the planet Earth as part of one large organism.

Figure 1.1 *All the plants and animals in this forest are dependent upon each other. They work together as an ecosystem.*

Look at the picture of a forest in Tobago. It is an **ecosystem**; it consists of an environment and a community of interacting plants and animals. *Eco* comes from the Greek word meaning *house*; *system* means *a group of things that interact with one another*. The many different organisms in an ecosystem – plants, birds, insects and mammals – are dependent upon one another (directly or indirectly) for their survival. For example, moss grows on the bark of a tree, the hummingbird draws nectar from the heliconia flowers and pollinates them in return, and the tiger cat preys on the lizard. The study of the **interactions** between organisms and their environment is called **ecology**, and so ecology is the study of ecosystems.

An ecosystem is made up of living and non-living parts

Ecologists divide an ecosystem into its **biotic** (living) parts, such as plants, animals, insects and micro-organisms; and its **abiotic** (physical or non-living) parts, such as light, temperature, water, gases, wind and the soil. An abiotic component such as water can vary in different environments. Look at the examples shown on page 7 (Figures 1.2a–d).

The abiotic parts of a particular place determine which plants can grow, and which animals can live there. When species of plants and animals adapt to the physical conditions of their environment, we call this process **evolution**.

Science Extra

Abiotic components also include features of the landscape such as altitude and the steepness of a slope.

Figure 1.2a *Very little water in a desert makes a good home for this cactus, which can survive in drought-like conditions.*

Figure 1.2b *Plenty of water in this swamp makes a good home for a mangrove tree.*

Figure 1.2c *The fresh water in this pond is the natural environment of this frog.*

Figure 1.2d *The salt water of the sea (with 3.5% average salinity) is the natural environment of these barracudas.*

An ecosystem in a bottle

An ecosystem is a sustainable self-contained unit. This means that it can continue for a long time on its own without needing anything from outside itself. A **terrarium** (a glass container in which plants are grown) is a model of an ecosystem. An **aquarium** (a tank in which fish and water plants are kept) is also a model of an ecosystem.

Figure 1.3 *A terrarium.*

ACTIVITIES

1. Write a short paragraph of your understanding of what an ecosystem is.
2. It is quite dark underneath the canopy of a forest.
   a. Describe two ways in which plants can maximise the sunlight they receive.
   b. List two other abiotic components and describe their particular quality in a forest.
3. Look at the picture of the terrarium.
   a. Where do plants in a terrarium get their water from?
   b. What would happen if you kept the terrarium inside a dark cupboard?

Assessment

Do some research on mangrove trees such as those that grow in the Nariva Swamp. What physical conditions do they need? Consider things like the salinity of the water and the acidity of the soil.

OR

Do a project in which you design an aquarium as a model of an ecosystem. Explain the role of each component of the aquarium.

2

Habitat – the place where plants grow and animals live

Keywords
community, habitat

Did you know?

The macaw lives in the Nariva Swamp and eats the fruit of only one type of palm tree.

Trinidad and Tobago and different types of habitat

Ecologists call the part of an ecosystem where a particular organism lives, its **habitat**.

Trinidad and Tobago has a rich and varied environment with many different habitats. Each habitat is characterised by the different kinds of plants growing there, and animals living there. For example, the golden tree frog is found in the montane forest habitat of the El Tucuche mountain area of the Northern Range.

Here are two more examples of different habitats.

Figure 2.1 *The golden tree frog.*

Figure 2.2 *The macaw.*

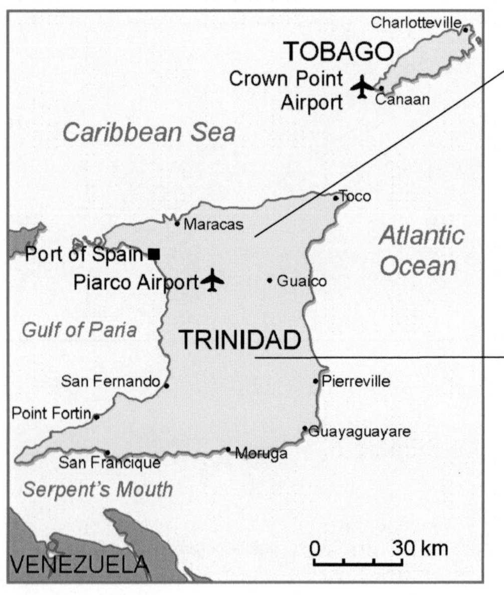

Figure 2.3 *This moist forest is characterised by tall evergreen trees, mora trees, vines, orchids, mosses and ferns. It is also home to numerous birds and butterfly species, venomous snakes, the howler monkey and the pig-like peccary.*

Figure 2.4 *The savannah area has sandy, acidic soil with a low nitrogen content. Plant species include grasses, the Moriche palm, and the insect-eating sundew plant.*

Animals and their habitat

The diet and lifestyle of an organism are suited to its habitat. Within a particular habitat, different animals live in different places, such as under a stone, or in a nest, or in a cave.

The Northern Range of Trinidad is characterised by montane forest and cliffs. The oilbirds or *diablotin* (French for *little devil*) are nocturnal and roost in the caves during the day. At night, they feed on the fruit of the oil palm and tropical laurels growing in this region.

Figure 2.5 *The habitat of these oilbirds is the dark Aripo Caves in the Northern Range of Trinidad.*

Habitat and community

The habitat for the oilbird is the dark caves in the Northern Range, but they are not the only animals that live there. The caves are also home to different species of bat and a species of blind catfish that lives in the water at the bottom of the caves.

We call the different species that live together in one habitat, its **community**.

Figure 2.6 *This bat is part of the oilbird's community.*

ACTIVITY

In groups, research different habitats in Trinidad and Tobago. Each group should give a five-minute presentation on what the physical environment is like, and what plants and animals live there.

Science Extra

Sulphurous pools are poisonous for most organisms, but some bacteria and algae have adapted to exist in them and nowhere else. Micro-organisms that live in harsh environments (at high pressure, high salinity or high temperature) are called extremophiles.

Assessment

The Buccoo Reef on Tobago provides a habitat for a whole variety of organisms such as sea anemones and sea sponges. Research and write a report or prepare a poster on this important habitat.

3

Relationships – plant and animal interactions

Keywords

carrying capacity, commensalism, competition, mutualism, parasitism, population, predation, symbiosis

Did you know?

The Azteca ants depend on the Cecropia trees (in Trinidad's Northern Range) for nest sites and food, while the trees depend on the ants for protection from leaf-eating herbivores and invasion by strangling vines. This is an example of **mutualism**.

Plant and animal interactions

Depending on the habitat, the community might consist of a large number of different plant and animal species living together. The interactions between different species within a community include:

- **competition** – this occurs when there is a limited supply of resources such as food, light or territory
- **predation** – when one organism (the predator) hunts and kills another organism (the prey)
- **symbiosis** – this is a close relationship between organisms of two different species (see below).

Symbiotic relationships

There are three types of symbiotic relationship:

Parasitism – the parasite lives on or in the body of another organism which is called the host. The parasite harms the host in some way. For example, ticks feed on the blood of many different mammals; this is an example of an animal-animal interaction.

Mutualism – a relationship between organisms of different species in which both organisms benefit and depend on each other. For example, bacteria live in the gut of termites and help them digest their food, getting their own food in return.

Commensalism – a relationship in which one organism benefits while the other is not affected in any way. For example, orchids grow on the high branches of tall trees. In this way they get sunlight, but they do not take away or give anything to the tree. This is an example of a plant-plant interaction.

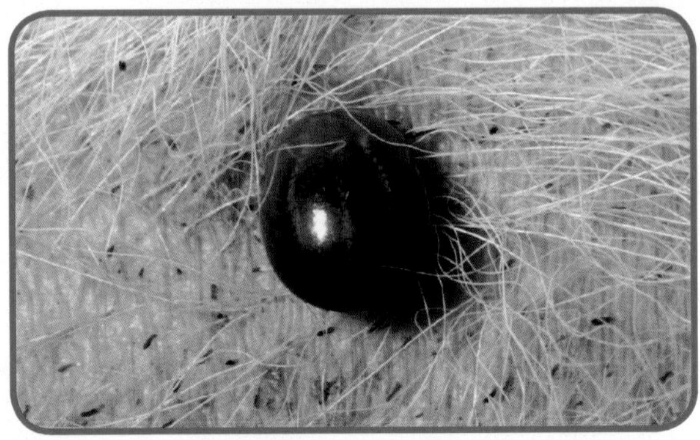

Figure 3.1 *This tick's body is swollen with the blood it has sucked from its host – a dog. This is an example of parasitism.*

Figure 3.2 *This is an epiphytic orchid – it grows on the tree but takes its nutrients from the air. This is an example of commensalism.*

Populations of plants and animals

You have seen that within an ecosystem community, there are different species of organisms. Each species or particular type of organism exists in the community as a population. A **population** is a group of the same species (organisms that interbreed) living in the same place or habitat.

There is a limit to the size of the population that a habitat will support. Most populations are stable in size. This is because the number of individuals in a population is not controlled by breeding potential, but by the environment, since all organisms need resources such as space and food.

If we plot a graph of the pattern of growth of a population (the increase in the number of individuals over time) we get an S-shaped curve like the one alongside.

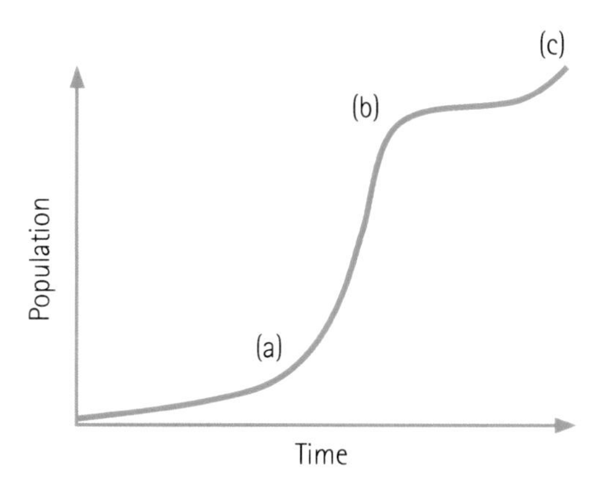

Figure 3.3 *An S-shaped growth curve.*

In part (a) of the curve, the population increases rapidly, but as the number of individuals increases, resources become limited, and so the growth rate slows down, as shown in part (b). Finally, the population stabilises or reaches equilibrium. This means that the rate at which individuals are born or enter the population, equals the rate at which others die or leave the population. There may however be small fluctuations, as shown in part (c). These fluctuations may be due to a favourable change in climate or a sudden outbreak of disease.

We call the maximum population that a habitat will support, its **carrying capacity**. For animal populations, the carrying capacity often depends on the food supply. For plants, the availability of water and access to sunlight are limiting factors. Both plants and animals compete for space or territory.

ACTIVITY

Use the picture of the forest ecosystem on page 6 to help you briefly describe and classify two:

- plant-plant interactions
- plant-animal interactions
- animal-animal interactions.

Science Extra

Competition is usually greatest at the same feeding level. For example, Indian mussels in the Caroni Basin breed faster than the indigenous mussels, and so are threatening the indigenous species. As the population of the Indian mussel increases, the population of the indigenous mussel declines.

Assessment

Name and explain three factors that limit the population growth of an organism. Research an example of a local plant or animal species whose population number is declining. What are the factors responsible for the decline?

4

Food chains, food webs and energy flow

Keywords

food chain, food web, pyramid of numbers

Did you know?

The feeding levels in a food chain are also known as trophic levels. The Greek word *trophos* means *feeder*.

Food chains – simple feeding relationships

Ecologists study the flow of energy and nutrients through an ecosystem by looking at what eats what or who eats who. The Sun's energy, which is captured by plants, passes from one organism to another along a food chain. Energy is lost by the organisms in the form of heat, and in the transfer process.

In a **food chain**, organisms are ordered by their relationship to each other as food or feeder, prey or predator. In the diagram of a food chain below, the froghopper insect eats the sugar cane, and the froghopper is in turn eaten by the keskidee bird. The arrows are drawn to show the direction in which the energy flows.

Figure 4.1 *An example of a food chain.*

The pyramid of numbers

A food chain tells us what eats what, but not how much is being eaten. If we want to illustrate this, we can show this using a **pyramid of numbers**. Let's say that 1 000 stalks of sugar cane are eaten by 500 froghoppers, which are then eaten by 20 keskidees. The width of each feeding level on the pyramid corresponds to the number of organisms eaten. Sometimes, instead of numbers, the pyramid shows the weight of food eaten, e.g. 1 000 tonnes of phytoplankton support one tonne of shark in the food chain.

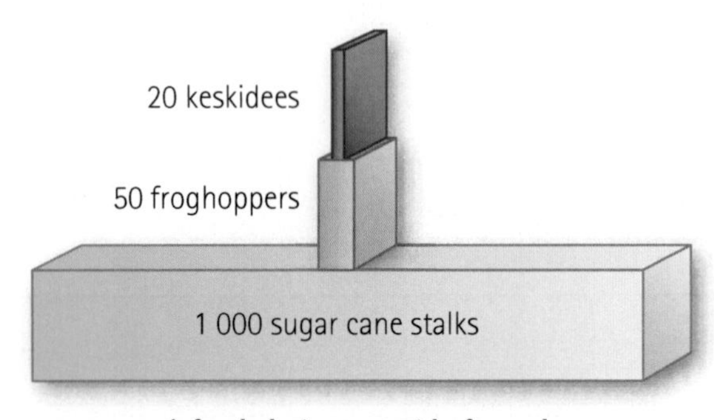

Figure 4.2 *A food chain pyramid of numbers.*

Food webs show more complex feeding relationships

As most animals eat more than one type of food, a food chain does not show all the feeding relationships. Food chains can be linked together as **food webs**, like the one shown alongside. Each different ecosystem will have a different food web, because each ecosystem has a different community of organisms.

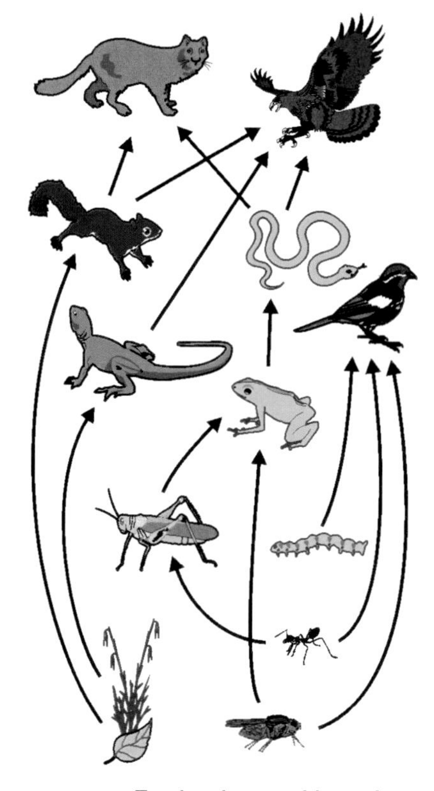

Figure 4.3 *Food webs are able to show more complex feeding relationships than food chains.*

Food webs can be disrupted

Toxic substances such as heavy metals and pesticides can accumulate in the food chain. Unlike food, these substances remain in the organism that eats them. Whatever happens to the population of one member of the food web is likely to affect the others. Food webs can therefore serve as useful tools to predict what might happen if a link in the food chain is disrupted. On the graph you can see the correlation between the lynx (predator) population and the hare (prey) population.

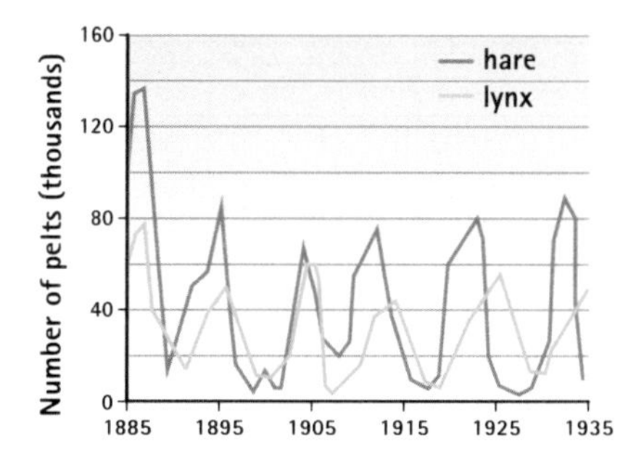

Figure 4.4 *Graph of hare and lynx populations.*

ACTIVITIES

1 Look at the food chain pyramid on page 12 and use it to answer the following questions:
   a In which direction does the energy flow?
   b How many feeding levels are there?
   c How many links (links are equivalent to arrows) are there between the feeding levels?
   d The food chains of most top predators usually have only three links (four levels). Your challenge is to come up with a food chain example consisting of five or more links.

2 Identify and write out three different food chains in the food web on the right. Which species would be affected if the population numbers of the wild cat decreased suddenly, owing to hunting by humans?

3 Illustrate for yourself how a pesticide like DDT can be concentrated at the top of the food chain by:
   - drawing in your book an empty pyramid of numbers with four feeding levels
   - writing on the following labels in the right order: grasshopper, plants, eagle, snake
   - making 20 evenly spaced pencil dots in each of the levels to represent the DDT residue.

   Do you think accumulation is more of a problem in short or long food chains? Justify your answer.

Assessment

Research and draw a food web for a habitat of your choice in Trinidad or Tobago.

5 The food chain – plants as producers, animals as consumers

Keywords

autotroph, carbohydrates, carnivore, consumer, herbivore, heterotroph, omnivore, photosynthesis, primary consumer, producer, secondary consumer

Did you know?

Ninety-nine per cent of Earth's organic matter is made up of plants and algae.

Plants produce carbohydrates as food

Figure 5.1 *The leaves of plants are the solar panels of nature.*

Science Extra

The first photosynthetic organisms, which were bacteria, evolved about 3.5 billion years ago. Like plants, they contain photosynthetic pigments such as chlorophyll.

In an ecosystem, the flow of energy and nutrients begins with a process called **photosynthesis**. The leaves of plants are like small, natural solar panels that trap about 1% of the Sun's energy. In Year 1 you learnt that *photosynthesis* means *building with light*. Plants use the energy of the Sun to make **carbohydrates**.

Because plants produce their own food, they are classified as **producers** or **autotrophs**, meaning *self-feeders*. They produce food not only for themselves, but also for animals. Through photosynthesis, 120 billion tonnes (metric) of new plant material are produced every year. You will learn more about photosynthesis in Unit 8 and about the carbon cycle in Unit 22.

All animals are consumers

All animals in the food chain are classified as **consumers**, because they cannot produce their own food. Consumers are also called **heterotrophs**, meaning that they feed on others. Many animals, like the manatee, only eat plants. We call these animals **primary consumers** or **herbivores**, meaning *plant eaters*. A snail is a herbivore, and so is an elephant. Each ecosystem has its own characteristic set of herbivores. Some animals, such as the tiger cat (ocelot), eat only other animals – birds, rodents, reptiles and insects. We call these animal-eating animals **secondary consumers** or **carnivores**, meaning *meat eaters*. Some animals, such as most humans, eat both plants and other animals. These animals are also secondary consumers, and are called **omnivores**, meaning *all eaters*.

Figure 5.2 *The gentle and graceful manatee of the Nariva Swamp is a primary consumer or herbivore.*

Figure 5.3 *This tiger cat (ocelot) is a secondary consumer or carnivore.*

ACTIVITIES

1 Answer the following:
   a Explain why plants are called *producers*. What is another name for organisms that make their own food?
   b List five animals that are herbivores.
   c Classify each of the following animals as a primary or secondary consumer, and if it is a secondary consumer, whether it is a carnivore or an omnivore: killer whale, rat, bat, and Scarlet Ibis.

2 Although most humans are omnivorous, they can easily survive on plant food only, providing they eat a good diet. A meat-eating diet requires seven times more agricultural land to feed one person than a vegetarian diet. Why is this? Hold a class discussion about vegetarianism. Why do people choose to be vegetarian? What are the benefits and drawbacks for the vegetarian? What are the benefits and drawbacks for the environment?

Assessment

Make a list of the plant foods, excluding fresh fruit and vegetables, that you eat.

6

Nature's decomposers

Keywords

bacteria, decomposer, detritivore, fungi, hyphae, micro-organism, spore

Did you know?

The spores of different kinds of fungi are often different colours, which is why mould can be green or black, or even bright orange.

The organisms that feed on the refuse or dead food in an ecosystem – dung, carcasses, dead leaves, fallen logs, even the discarded exoskeletons of insects – are called **detritivores**. They range from the largest scavengers such as vultures and crabs, to the smallest decomposers such as fungi and bacteria.

Fungi and bacteria

Fungi and bacteria play an essential role as decomposers in the ecosystem. As soon as plants and animals die, they get to work, breaking down the organic matter. Generally, these organisms are at the end of the food chain, but some mushroom fungi are edible, and will themselves be on the menu in the food web.

Fungi reproduce by means of **spores**. The tiny spores of the fungus are present everywhere in the air and the soil. Under favourable conditions, the spores germinate, forming threadlike **hyphae** (from the Greek word, *hyphe*, meaning *web*). The mass of hyphae make up the mycelium. Some of the hyphae work like the roots of a plant to anchor the bulk of the fungus and absorb nutrients. They also secrete digestive enzymes. The aerial hyphae work like stalks, producing the fruiting bodies that carry the spores.

Figure 6.1 *Two kinds of fungi: toadstools (above) and bracket fungi (below).*

Bacteria are single-celled decomposers. They are only visible under a microscope and so they are called **micro-organisms**. Different groups of bacteria play different but specific roles as decomposers – some specialise in digesting cellulose, others proteins, and others carbohydrates. Many of them make use of special enzymes to digest the organic compounds that other animals are unable to. While you get your carbon and nitrogen from meat and potatoes, some bacteria can get theirs by degrading fur, claws or feathers.

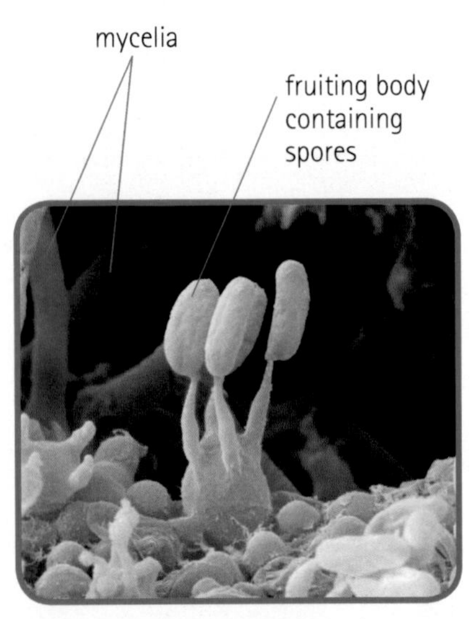

Figure 6.2 *A magnified view of a fungus.*

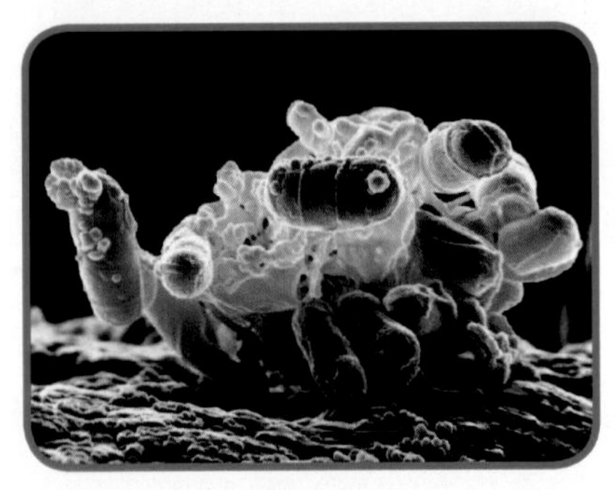

Figure 6.3 *A cluster of bacteria. The individual bacteria in this picture are oblong and brown coloured.*

The decomposers release nutrients back into the soil

Dozens of elements form thousands of different complex molecules that join together to make a living organism. **Decomposers** break down these complex molecules into simple ones, releasing them back into the soil, and making them available to plants. Although plants make their own food by photosynthesis, they need supplementary nutrients for good growth. Plants in turn make these nutrients available to animals, and in this way, the nutrients in an ecosystem are recycled.

What happens to the energy?

At the end of the chain in a food web, the energy that was originally captured by plants from the Sun is lost as heat. Some of the energy that bacteria get from their food source is used for themselves, and the rest is released as heat.

Have you ever noticed how the inside of a compost heap gets hot from all that microbial activity?

What conditions do fungi such as bread mould like?

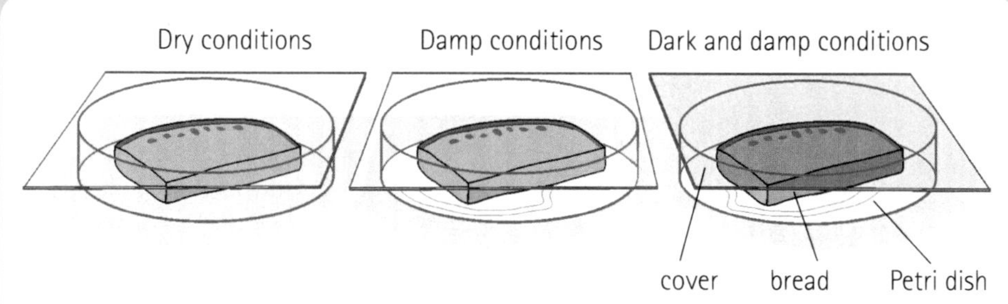

Figure 6.4 *Investigation for bread mould.*

1. Set up three covered Petri dishes (or plastic bags) under the three different conditions shown in the diagram. Place a piece of bread in each. Leave for a few days.
2. Record which samples the bread mould has grown on.
3. What do you conclude are the best conditions for the fungus to grow in?
4. Draw a detailed sketch of the bread mould. Label the threadlike hyphae and the darker spores. (If possible, take a closer look at it under a microscope.)

Safety: Make sure not to touch the mould, and remember to wash your hands well afterwards. Throw the pieces of bread away in a sealed bag after you have done your experiment.

Energy from decay

The energy stored in decaying organic matter can be used as fuel. A biogas digester is a large tank that stores animal dung. As the bacteria in the dung break it down, they produce methane gas that collects in the tank, ready for use for cooking and heating.

Figure 6.5 *Biogas digesters can produce enough gas to run a household.*

Assessment

What would happen if there were no decomposers to break down the bodies of organisms that were once alive? Write an interesting scenario. OR write on the following topic: 'In nature, nothing goes to waste.'

7

Nature's scavengers

Keywords

scavenger

Did you know?

In a forest ecosystem, more than 90% of the plant matter (biomass) is ultimately consumed by detritivores, and not herbivores.

The scavengers: waste disposal and soil care

Scavengers such as the turkey vulture feed on the carcasses of dead animals. The bird urinates on its legs and, because the urine is very concentrated, it acts as a disinfectant, killing the micro-organisms that the bird picks up from carrion.

Dung beetles are recyclers in many ecosystems since they live in a range of habitats – desert, grassland, forest and farmland. The eggs are laid in the ball of dung and the hatched larvae feed on the dung's undigested plant fibre. The adult beetles feed only on the liquid part of the dung, by squeezing and sucking out the juices. Other scavengers feed on dead, rotting plant and animal matter in the soil. These scavengers include earthworms, beetles, woodlice, millipedes and termites.

Figure 7.1 *A turkey vulture eating a road-killed snake.*

Figure 7.2 *The dung beetle gets all its nutrients from dung.*

Figure 7.3 *Earthworms aerate the soil.*

Earthworms and termites for healthy soil in the ecosystem

As you have learnt, soil is an abiotic component of ecosystems. Scavengers such as earthworms and termites play an important role in soil health. Earthworms burrow through the soil, feeding on humus. In the process, they aerate and mix the soil, carrying dead organic matter deeper into the soil. Without the feeding activity of earthworms, the soil would be hard and airless.

Termites live in a variety of habitats – mangrove, wetlands, forest and grasslands – and play an important role in recycling fallen trees back into the soil. (They can also chomp their way through furniture made of indigenous wood!) They have special bacteria in their gut to help them digest tough, cellulose-rich plant material such as wood and grass. Termites are social insects – there are more than 100 000 in one nest, which contains soldiers, workers and reproductives. The soldiers protect the nest; the workers build the nest, forage for food, feed the soldiers and reproductives, and remove the bodies of dead or dying termites. In making their nests or mounds, they dig up rich soil from underground and bring it to the surface. The reproductives mate and lay eggs.

Figure 7.4 *Nasute termites searching for food.*

INVESTIGATION Identifying scavengers in the soil

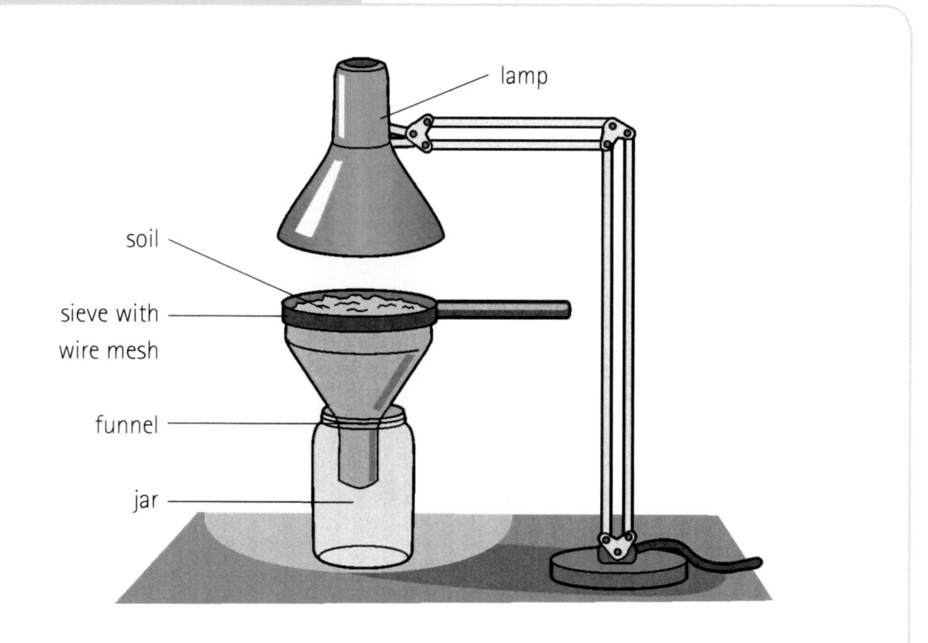

Figure 7.5 *Apparatus for identifying scavengers in the soil.*

Collect a crumbly soil sample rich in organic matter from a shady, leafy place. Set up the apparatus as shown above and switch on the lamp. After one hour, inspect the bottom of the jar for small creatures or insects. Gently transfer them into the lid of the jar and take a closer look.

1 Identify the organisms, using reference books if necessary.
2 Make labelled sketches of the organisms.
3 Explain why a sample of dry, sandy soil is unlikely to contain many scavengers.
4 Consider the habitat of soil organisms – what sort of conditions do they like?
5 Can you explain the use of the lamp in this investigation?

Assessment

Find out more about the role of earthworms, termites or a different scavenger in the ecosystem. Now write a short article, as if for a nature magazine.

8

Respiration and photosynthesis – oxygen, carbon and energy cycling

Keywords

carbon dioxide, cellular respiration, chlorophyll, chloroplast, glucose, mitochondria, oxygen, photosynthesis

Plants and animals get their energy from food by cellular respiration

Getting energy from food is a bit like burning a piece of wood. The wood contains the energy, but we need to set fire to it to release the energy for warmth. We also need to 'burn' the food we eat to release energy. The process of getting energy from food for growth, warmth, metabolic processes, and activity or movement is called **cellular respiration**. As the name suggests, cellular respiration takes place in the cells of living organisms. Just like animals, plants also respire. They too need to get energy from the food they make.

Food comes in the form of carbohydrates (starch and complex sugars), proteins, or fats. Carbohydrates must first be broken down and converted to **glucose**, which is a simple sugar. Glucose is the reactant or starting ingredient of cellular respiration.

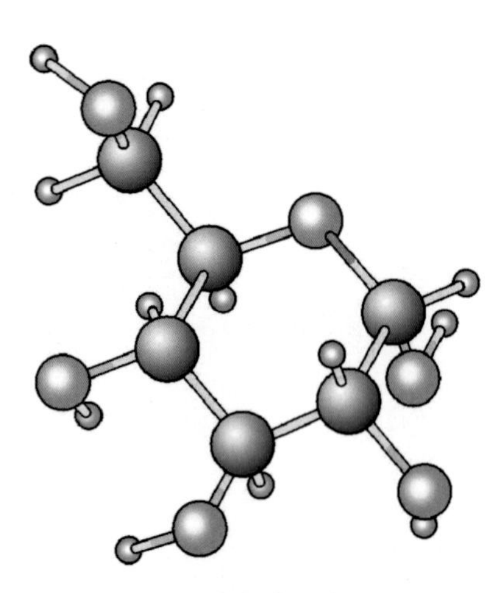

Figure 8.1 *A model of a glucose molecule. The carbon atoms are grey, the oxygen atoms are brown, and the hydrogen atoms are green.*

The respiration reaction

Respiration takes place in cell organelles called **mitochondria** (singular: mitochondrion). The name is derived from the Greek words, *mitos* meaning *thread* and *chondrion* meaning *grain*.

Do you agree that the mitochondrion looks like a small grain with threads inside?

In the overall cellular respiration reaction, glucose combines with **oxygen** to release energy. The by-products of this reaction are **carbon dioxide** and water. (Remember how in Year 1 you tested for these products that you breathe out through your lungs?)

glucose + oxygen → carbon dioxide + water + energy

$C_6H_{12}O_6 + 6O_2 \rightarrow 6CO_2 + 6H_2O + \text{energy}$

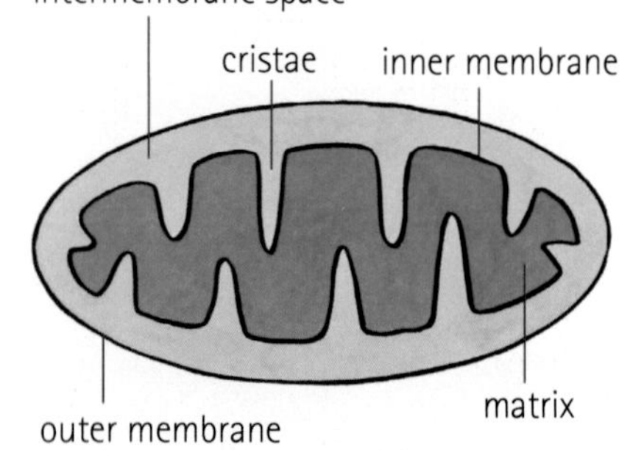

Figure 8.2 *The mitochondrion is where respiration takes place.*

Plants trap energy from the Sun by **photosynthesis** to make food and fuel. In Year 1 you saw how to test a leaf for starch using iodine as an indicator. Starch is the main food storage product in plants – think of plant foods such as rice, potatoes and plantains. Plants can also convert the glucose produced by photosynthesis (see opposite) into other food groups, such as protein (in beans), sugars (in fruits), and oils (in nuts and seeds). In plant-based fuels such as wood and fossil fuels, energy is stored in other hydrocarbon molecules.

The photosynthesis reaction

The **chloroplasts** in plant cells are the site of the photosynthesis reaction. *Chloro* means *green*, because the chloroplasts contain the green **chlorophyll** pigment. Here, the raw materials – water and carbon dioxide – that the plant has taken up are combined with the energy of the Sun to produce glucose, with oxygen being released in the process.

$$\text{carbon dioxide} + \text{water} + \frac{\text{sunlight (energy)}}{\text{chlorophyll}} \rightarrow \text{glucose} + \text{oxygen}$$

$$6CO_2 + 6H_2O + \frac{\text{sunlight (energy)}}{\text{chlorophyll}} \rightarrow C_6H_{12}O_6 + 6O_2$$

Figure 8.3 *Photosynthesis takes place in the chloroplasts in leaf cells. This drawing shows a magnified chloroplast. Each leaf cell contains many chloroplasts.*

Photosynthesis and respiration are opposite processes

Take another look at the respiration and photosynthesis reactions. In the one reaction, energy is absorbed, in the other it is released. And the products of one reaction are the reactants of the other.

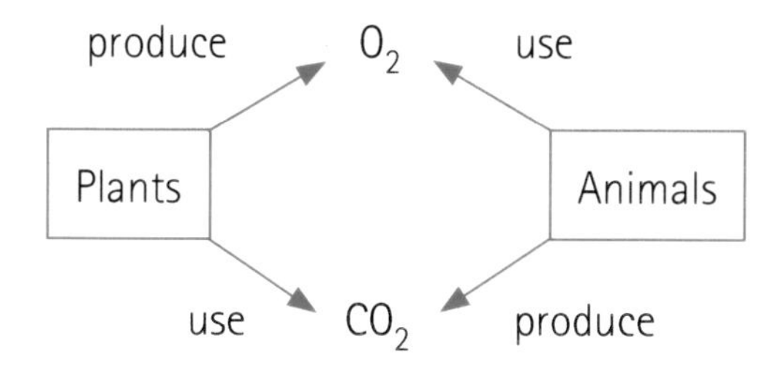

Can you see that respiration is essentially the reverse process of photosynthesis?

The balance of carbon dioxide and oxygen

Plants produce oxygen through photosynthesis, and thus trees are sometimes called 'the lungs of the Earth'. The natural processes of photosynthesis and respiration are in balance. But with increased human activity, such as the burning of fossil fuels (plant-derived products) and the destruction of forests, the amount of carbon dioxide released by combustion and respiration is now more than the amount taken up by plants by photosynthesis. (See the carbon cycle, Unit 22, and the greenhouse effect and global warming, Unit 32.)

ACTIVITY

Explain why respiration and photosynthesis can be thought of as opposite processes. What are the products of respiration? What are the reactants of photosynthesis?

Assessment

Explain this bumper sticker: 'Have you thanked a green plant today?'.

9

Plants as habitat and resource

Keywords

epiphyte, indigenous, traditional medicine

Did you know?

About 27 500 different plant species have been identified on Earth. Only about 10% of these have been investigated for their value to humans.

In Unit 2 you looked briefly at how different habitats are characterised by different vegetation. Natural habitats contain **indigenous** plant species, whereas human-made habitats, such as agricultural fields, usually contain introduced plants. Here we will look in more detail at plants, both as a habitat and as a resource for humans.

The vegetation of Trinidad and Tobago

The pie chart shows the proportions of the different ecosystem areas in Trinidad and Tobago relative to the total land area. This data is based mainly on satellite images recorded by the Global Land Cover Characteristics (GLCC) project.

Forest is the dominant habitat of the islands, consisting of tall evergreen or deciduous trees that make the upper canopy. Besides trees, the forests are well populated with ferns, lianas and epiphytes. Liana is a general name for tangled woody vines or climbers of different species. They use trees to reach the top of the canopy and from there, spread from tree to tree, providing animals of the forest with travel routes as they serve as ladders, ropes and bridges. **Epiphytes** include plants such as some mosses, orchids and bromeliads. These plants grow on, or attach to other plants – usually only for physical support.

Figure 9.1 *The indigenous Chaconia of Trinidad and Tobago is our national flower.*

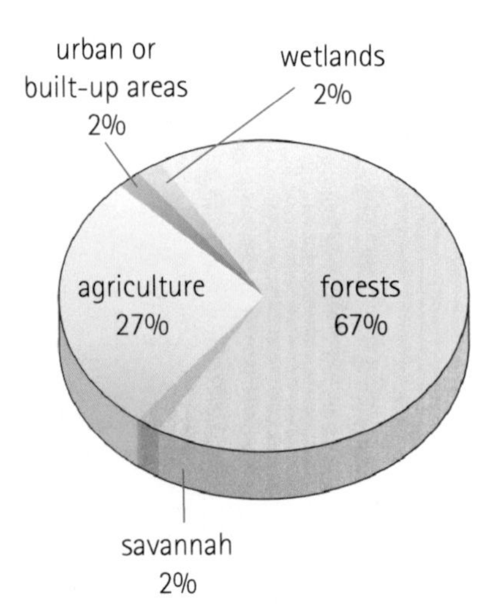

Figure 9.2 *The proportions of the different ecosystem areas.*

Plants for agriculture and industry

Although not indigenous, sugar cane is widely cultivated in the Caribbean and is an important economic crop. All over Trinidad and Tobago you will see sugar cane fields and sugar refineries that process the cane. The sugar cane industry is very important for our country's economy, while Trinidadian rum is famous all over the world. Other important food crops include rice, coconut, coffee and citrus fruit. Trees are an important source of timber, firewood and paper pulp.

Figure 9.3 *A sugar cane field.*

Science Extra

Many crop plants are plants that have been domesticated and improved by breeding and selection. Maize, one of the most common crops, was bred from a wild South American grass called teosinte 7 000 years ago.

Plants for traditional medicine

About 150 different plants are used as **traditional medicine** or herbal remedies. Tea brewed from the yellow leaf of breadfruit is used to treat diabetes and high blood pressure. Fruit from the Bilimbi or cucumber tree is used as a natural remedy for digestive ailments.

Figure 9.4 *A cucumber tree. This tree is a member of the gourd family, which includes melons, cucumbers and pumpkins.*

ACTIVITIES

1. Give a brief description of the characteristics of forest, savannah, cropland and wetlands, with four examples of specific plants that grow in each of these different habitats. To describe cropland, list the five main crops in Trinidad and Tobago.
2. Find out which other industries in Trinidad and Tobago make use of plants.
3. Knowledge about traditional medicine is dying out. Interview an older person about some of the uses of plants that are no longer common today.

Assessment

Do you know the names of the common trees of Trinidad and Tobago? Can you identify them? Make an identification poster for the common trees in your area. Do a tree survey. How many of each type of tree can you find in your neighbourhood?

Studying an ecosystem

Keywords

transect survey

Did you know?

The GLOBE project was set up by NASA to encourage students from around the world to investigate their environment. You might like to do this investigation as part of the GLOBE project and then publish your results on the Internet for other students to look at. Look on www.globe.gov or www.globe.org.uk for more details.

This is now your chance to apply what you've learnt about ecosystems by going outside and studying your environment. Choose an area near to your school or home to investigate.

INVESTIGATION Study of an ecosystem

1 Choose an ecosystem to study. It could be a pond, a tree, the whole schoolyard or your back garden.
2 Draw a map of your ecosystem area showing the main features, such as buildings, fences, trees, and orientation relative to the Sun (see the sketch map below). You could also include a photograph.
3 Look at the previous units to give you some ideas for your investigation. Do you want to count the plant and animal populations, for example, or explore the feeding relationships in your ecosystem? Perhaps you want to ask and answer a research question such as, 'Is my backyard a healthy ecosystem?', or 'How does the population of millipedes vary between January and March?'.
4 Collect your data by a combination of measurement and observation. Look and listen for indirect signs of animals, such as tracks, holes in the ground, moulting, dung or droppings, and bird-calls. Also look for signs of feeding activity such as holes in leaves or fruits, and scratching or digging marks in the ground. You might like to use a particular method such as a transect survey, or make use of a trap to catch insects (see the instructions on page 25).
5 Present your results as a poster or a report. Include the following:
   - Title – to say what your project is about; ideally it should be short and interesting.
   - Introduction – what question(s) were you asking? It should be short and clear.
   - Method(s) – how did you collect your data?
   - Results – present your findings as tables, graphs, diagrams and photographs. Give these captions.
   - Conclusion – what have you learnt from this study?
   - References – include a list of sources you have used, e.g. books, magazine articles, the Internet.

Figure 10.1 *An example of a sketch map of the area you are going to study.*

A pitfall trap

You can make a simple trap to collect invertebrates such as insects. It is especially useful for collecting nocturnal insects.

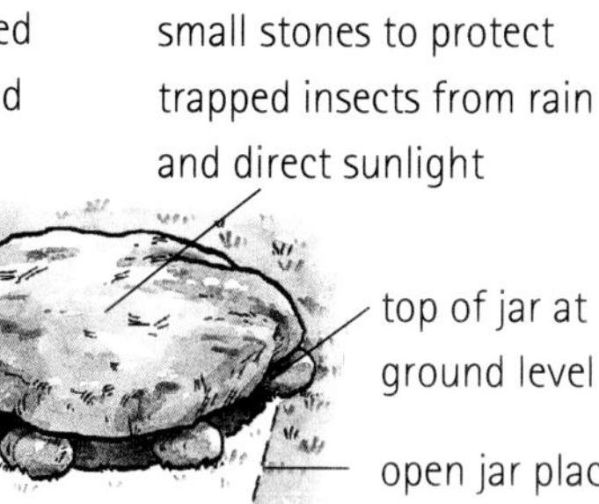

Figure 10.2 *A pitfall trap to collect invertebrates such as insects.*

A transect survey

Choose an area, with some variation in the plants, animals and abiotic conditions, that covers a 10 m distance. For this **transect survey** you will need two pieces of wood to use as pegs, and a long piece of string. Tie each end of the string to a peg. Now stretch the string out in a line where you want to do the survey. Mark each metre along the length. Mark on your sketch map where your transect is. Now begin at one end of the string and move along the length. Record the abiotic conditions and the number and types of plant and animal species you find touching the string within each metre length. Record your results in a table like this:

Distance (metres)	Animals – types and numbers	Plants – types and numbers	Abiotic conditions (water, light, etc.)
0–1 m			
1–2 m			

Figure 10.3 *A transect survey is a way of collecting data along a strip of land.*

Assessment

What did you learn from your ecosystem investigation? Did you enjoy the investigation? Give reasons for your answer. Were there any parts of the investigation or presentation you could have done better?

Endangered and extinct species

Keywords

endangered species, endemic, extinct, loss of habitat

Did you know?

Scientists estimate that worldwide, about 100 species become extinct every day.

Science Extra

The percentage of species that are endemic to Trinidad and Tobago is less than in the rest of the Caribbean because our country shares many species with the South American mainland.

Species under threat in Trinidad and Tobago

Trinidad and Tobago has a great diversity of plant and animal species. Of these, about 2% of the animals and 10% of the plants are **endemic**. This means that these species are found here and nowhere else. Island plants are easily threatened because land space is limited, and foreign plants can out-compete them. Research needs to be done to find out which of our particular species are rare or endangered.

Here are examples of three of our **endangered species**:

- The giant leatherback turtles are one of the wonders of Tobago. They are most easily seen in March when they come ashore to breed, attracting tourists and locals alike. Unfortunately, the number of turtles has declined drastically because of **loss of habitat** (particularly nesting ground), hunting, poaching of eggs, and getting trapped in fishing nets and lines. As a result, they are now an endangered species. The Save Our Sea Turtles (SOS) Tobago project aims to protect them.

Figure 11.1 *A female leatherback turtle digging her nest hole on a beach.*

- The tiger cat, also called the ocelot, has been hunted for its beautiful coat and has also suffered from loss of habitat. The Paria Springs Trust runs an Ocelot Conservation project that aims to conserve large areas of natural habitat for the wild cat. (You will see a picture of the ocelot on the front cover of this book.)
- The green iguana is classified as vulnerable on the Red Data List, but island populations such as those on Dominica and St Lucia are severely threatened. The giant lizards are hunted by foreign predators such as dogs, cats and rats, and have to compete for food with grazing farm animals such as sheep and goats.

Figure 11.2 *The green iguana is a threatened species.*

Protecting our endangered species

Loss of habitat threatens the survival of particular plant and animal species. Because habitats such as swamps and coral reefs are home to unique animals or plants, some are now protected, for example the Aripo Savannah and the Nariva and Caroni swamps. Creation of reserves such as the Caroni Bird Sanctuary offers protection to our national bird, the Scarlet Ibis.

Organisations such as the International Union for Conservation of Nature (IUCN) run a Red List data project that monitors the status of threatened plant and animal species (see www.iucnredlist.org). These include the aye-aye lemur of Madagascar, the Komodo dragon of Indonesia, the kakapo parrot of New Zealand, and the gorilla and white rhino in Zaire.

Figure 11.3 *The Scarlet Ibis – our national bird.*

Extinct species

If all members of a species die, we say that the species is **extinct**. The dodo, the sabre-toothed tiger, the mammoth and the dinosaurs are all well-known examples of extinct animals. What caused the extinction of the prehistoric dinosaurs is not clear. It could have been a catastrophic event such as the impact of a large meteor, causing extensive fires and smoke that blocked out the sunlight from Earth for a long period of time. The dodo of Mauritius, however, is an example of a recent extinction for which humans were responsible. Isolated on an island, these flightless birds were unafraid of humans, having never seen them before. The whole population was soon hunted to extinction by sailors who stopped off to collect food for their journey.

Figure 11.4 *The dodo bird – hunted to extinction.*

ACTIVITIES

1. Having read the examples above, list reasons why some species become endangered.
2. Write a one-page essay on why we should protect endangered species.
3. Draw a mind-map or spider-gram, showing the ways in which declining populations of species can be protected or re-established.
4. Bearing in mind that many prehistoric animals became extinct by natural processes, do humans have a moral obligation to preserve habitats and protect wildlife? Hold a class discussion as to who should be responsible for supporting endangered species.

Assessment

Consider the quagga or the thylacine, the Barbados raccoon, the broad-faced potoroo, or the Vietnam warty pig. Research a lesser-known animal that has become extinct within the last 500 years. What was its diet and habitat? When and why did it become extinct?

12

Deforestation

Keywords

deforestation, slash-and-burn agriculture, soil erosion

Did you know?

It is estimated that in South America, a forest area the size of Trinidad and Tobago is being cut down every four months.

Rainforests are some of the world's oldest ecosystems. They contain half of the world's plant and animal species. But how much longer will they last? **Deforestation** is the loss of forested areas and is one of the saddest aspects of environmental degradation. It results in the loss of habitat for resident plants and animals, and in the loss of trees, which use carbon dioxide and are part of the water cycle through transpiration.

Deforestation in Trinidad and Tobago

About half the land area (226 000 hectares) consists of forest. A hectare is roughly the size of two soccer fields. Between 1990 and 2000, the islands lost 4% of their forest cover.

The reasons for deforestation include logging for timber and firewood, forest clearance for industry and agriculture (crops and livestock), and forest fires.

Forest fires occur every year in our country, particularly in the Northern Range of the mountains of Trinidad.

Figure 12.1 *Forest fires cause a huge amount of damage.*

Slash-and-burn

Slash-and-burn agriculture is a method of clearing forests before crops are planted. After a few years the soil becomes poor, and the farmer moves to a new area. This type of agriculture is a traditional practice and is sustainable as long as it is done on a small scale and the forest is left long enough to recover. This can take several lifetimes.

However, in recent years this method has been practised illegally in protected areas and there is the added danger of runaway fires. Owing to improved conservation efforts, the rate of deforestation in Trinidad and Tobago has now dropped significantly. But uncontrolled slash-and-burn fires still cause most of our forest fires.

The impact of deforestation

Apart from the loss of habitat for plants and animals, the forest vegetation protects the soil from erosion. The tree canopy of a forest withstands most heavy rain, and the roots of the trees hold the soil together. Without trees, **soil erosion** takes place – the rain runs off quickly, carrying the fertile topsoil with it.

Science Extra

Forest fires are not bad for all plants. The fires clear the ground, giving plants space and light to grow. Some specialised plants have adapted to use the heat from the fire to germinate their seeds so that they emerge only at this time.

ACTIVITY

Read the extract alongside and answer the following.

1 List four human activities that are linked to forest fires in Indonesia.

2 Suggest ways of dealing with the problem.

The Northern Range mountains have recently been illegally deforested for agriculture and in subsequent years there has been flooding in the Caroni River area during the wet season. This area is fed from rivers in the Northern Range.

INVESTIGATION How does vegetation affect water runoff?

Try this as a simulation.

1 Make up two trays of sand, and cover one with vegetation as shown in the diagram.

2 Pour a litre of water over each tray, and collect the water that runs off each of them.

3 Observe the difference between the two trays and use your observations to explain how the deforestation in the Northern Range has led to flooding in Caroni.

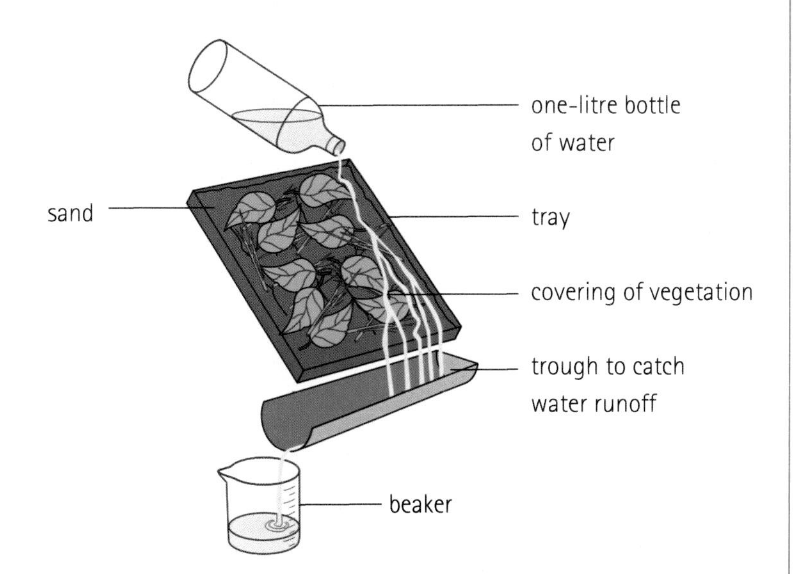

Figure 12.2 *Apparatus for investigating water runoff.*

Assessment

Design a poster to educate people about the dangers of forest fires or the impact of deforestation on the environment.

Study links logging with severity of forest fires

Tropical rainforests don't usually burn. In their natural state, fuel loads are low and not highly flammable, and the humidity is high even during drought years.

But Indonesia's rainforests have experienced the effects of heavy logging and slash-and-burn agriculture, resulting in weakened ecosystems. Indonesia's forests burned following droughts in 1982–83, 1987, 1991 and 1994.

Pressure on Indonesia's rainforests has been building for 30 years. Government relocation programmes encouraged people to move from densely populated regions to less populated islands such as Borneo. The increased population pressure has led to uncontrolled conversion of forest to agricultural use, which is done through slash-and-burn techniques.

Non-indigenous islanders have also used fire for hunting. For example, during the day, turtles usually stay in mud holes, where they are difficult to find. Fire forces them to come out, and then they can be easily collected.

Large areas of forest have also been cleared to make way for pulp wood and palm oil plantations. Much of the forest land that has been cleared is anchored in peat, which is a rich source of fuel for fires. Fires are also used as a 'weapon' in land disputes between the plantation companies and local people who consider the land to be theirs.

Unless land-use policies are changed to control logging and to introduce reduced-impact logging techniques, recurrent fires will lead to a complete loss of Borneo's lowland rainforests.

Source: National Geographic News, *3 December 2001 (adapted)*

13

The impact of pesticides on the environment

Keywords
bioaccumulation, biodegradable, biological control, biological control, monoculture, organic farming, pesticide

Apart from clearing land of its natural vegetation to make fields for crops and grazing, intensive or large-scale farming can have other negative effects on the environment.

Monoculture encourages pests

Sugar cane is the only crop grown in the Caroni area. This type of agriculture, in which only one crop is grown, is called **monoculture**.

Compared to a natural habitat such as a montane forest, a sugar cane field, created by humans, is a limited habitat.

Montane forest	Sugar cane field
Has a diversity of plant and animal species and complex food webs, which makes for more stable populations.	Has low diversity and simple food webs. This makes for unstable populations that are susceptible to changes in conditions, such as short-term drought.

Figure 13.1 *A montane forest in Trinidad and Tobago.*

The sugar cane habitat is ideal for the froghopper insect – hectare after hectare of sugar cane ready to eat, and no predators. The natural predators of the froghoppers are not adapted to the sugar cane field habitat, and so **pesticides** are used to control the insects.

Figure 13.2 *Crop-spraying with pesticides.*

Pests are controlled with pesticides

Some pesticides are **biodegradable**, meaning they are broken down in the environment once they have killed the insects, but some last as toxic residues. These pesticides can contaminate the soil, water sources, and the food chain. DDT is a pesticide that is now banned in most countries of the world. Insects that were killed by DDT were eaten by other animals, and the poison was passed up the food chain to birds of prey such as eagles. This process is called **bioaccumulation**. Because the poisoned birds laid thin-shelled eggs, their chicks failed to develop or hatch, and the eagle population declined. Another problem with pesticides is that the target insects can develop resistance.

Alternatives to pesticides: organic farming and do-it-yourself insect-resistant plants

Organic farming is a method of agriculture that does not use any pesticides, fungicides (chemicals that kill fungi) or artificial fertilisers. A combination of methods is used for pest control, e.g. companion planting and crop rotation, the use of insect traps and plant-derived pesticides, and the introduction of insect predators for **biological control**.

Another approach is to grow crop plants that have been genetically modified for resistance to insect pests. Crops such as Bt maize, Bt cotton and Bt soybean contain a gene from a bacterium called *Bacillus thuringiensis*, which produces a protein that is toxic to insects. Those in favour of genetically modified organisms (GMOs) argue that these crops are more environmentally friendly because they reduce the need to spray with pesticides and that the Bt toxin only kills the target pest so other insects and animals are not affected. On the other hand, people who are against GMOs are concerned that these foreign genes might be able to escape to other plants or organisms, that insects might develop resistance to the Bt toxin, or that the foreign insecticide protein in these crop plants might be harmful to humans.

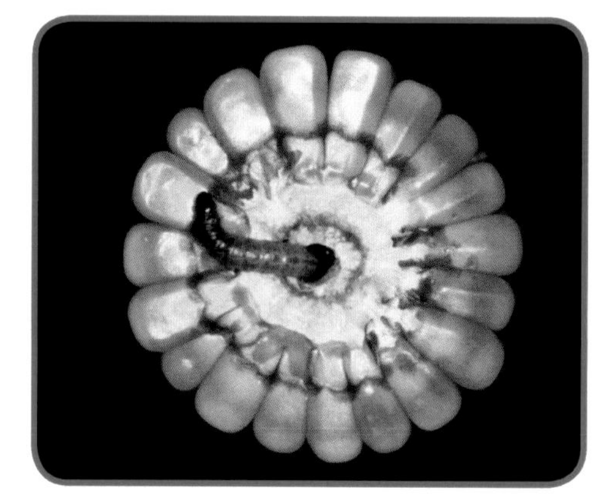

Figure 13.3 *The caterpillar of the corn borer moth, which damages maize crops by tunnelling into the cob.*

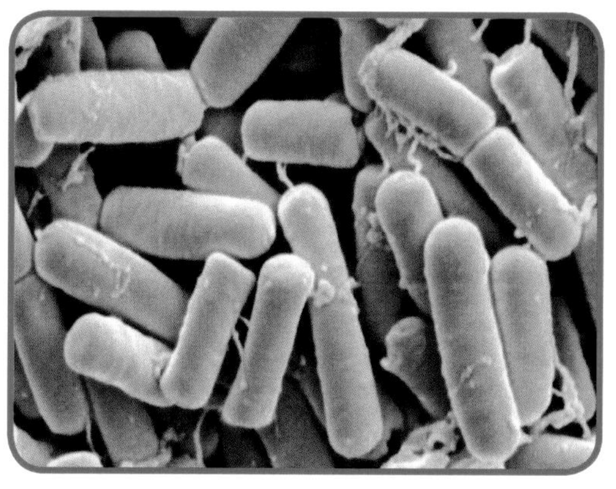

Figure 13.4 *Bacillus thuringiensis, which can be sprayed or dusted onto plants to protect them from pests.*

ACTIVITIES

1 Find out by means of interviews or research:
   a Which pesticides are commonly used in Trinidad and Tobago?
   b What measures are in place to ensure that crop foods are pesticide free?
   c What safety precautions do farmers take when spraying with pesticides?
   d What are the harmful effects of these pesticides, if eaten by humans?
2 Do some research on genetically modified organisms (GMOs). Have a class discussion about the advantages and disadvantages of genetically modified foods. Why not ask a farmer to join in with your discussion? Consider issues like food demands and patent rights. In what ways is the genetic engineering of crops the same or different from plant breeding?

Assessment

Write a summary of the class discussion for Activity 2 in your book.

Science Extra

How does the Bt crystalline protein kill the corn borer larva? When the borer eats the Bt crop plant, the toxic crystalline proteins make holes in the lining of the insect's gut.

14

Damage to aquatic ecosystems

Keywords

ecotourism, eutrophication, fertiliser, wetland

Did you know?

World Wetlands Day is celebrated on 2 February every year. Wetlands work so well at filtering pollutants that artificial wetlands are sometimes created for water purification and sewage treatment, such as the one created in Bon Accord.

Water is an important resource and a habitat

Water is a precious natural resource. Because rain falls freely, it is easy to forget that freshwater supplies must be looked after and conserved. But as the population grows and industry expands, the demand for water increases. Water tapped from boreholes can lower the level of the underground water table, causing saltwater seepage to take its place. And the redirection of water flow from rivers for human consumption can mean the loss of habitat for aquatic plants and animals. Not only do plants and animals need water for survival, but the water in rivers, lakes, wetlands, estuaries and the sea is a habitat for many species.

The harmful effect of fertilisers

Like pesticides, **fertilisers** can also have a harmful effect on ecosystems. Nitrates and phosphates in the fertilisers are washed into rivers, dams and the sea. This nutrient pollution encourages rampant growth of algae, which spread over the water surface, blocking out the light. This is called **eutrophication**. Some of these algal blooms are toxic to plants and animals, and the toxins can enter the human food chain through shellfish. As the algae die and decompose, bacteria use up oxygen from the water, causing fish and other animal life to suffocate from lack of oxygen.

Why wetlands should be protected

A **wetland** is an area of land covered with water, or an area of waterlogged land. It includes a wide range of habitats – shallow lakes, freshwater marshes, coastal lagoons, mangrove swamps, and floodplains. Trinidad and Tobago has many wetland ecosystems such as the Caroni and Nariva Swamps.

Figure 14.1 *The Nariva Swamp is a wetland ecosystem.*

Wetland ecosystems are important because they support a wide range of plants and animals, and often have complex food chains made up of fish, crabs, oysters and birds, which humans tap into. They also serve as nurseries for young fish, they act as water sponges by 'soaking up' floodwaters, and they filter pollutants from contaminated water by sedimentation. A sludge forms or settles on the bottom where it can be degraded by micro-organisms. Because of the diversity of wildlife, they are often sites for recreation and **ecotourism**.

The Nariva Swamp, the largest freshwater wetland of Trinidad and Tobago, is the habitat of the manatee, but is threatened by rice and watermelon cultivation and by the deforestation of its catchment area (i.e. the area where its source water falls as rain).

Figure 14.2 *A Caribbean spiny lobster crawling about on a coral reef.*

ACTIVITY

Read the extract about Buccoo Reef, and answer these questions in your own words.

1 Why are the coral reefs important to humans?
2 Why is plankton important in the food chain?
3 Name two kinds of coral.
4 What has caused damage to Buccoo Reef?

Assessment

Invent your own slogan to advertise how essential water is for plant and animal life. An example is, 'Water – more precious than gold'.

Tobago's reefs – fragile wonders that need your help

Coral reefs are not just beautiful natural wonders designed to provide people with a relaxing distraction while bobbing around on a boat or with a snorkel. They are a vital part of the ecological fabric and economic activities of small Caribbean islands. They are the backbone of Tobago's two largest industries – tourism and fishing – providing both jobs and food. They also protect the coastline from erosion by breaking ocean swells. They produce the sand on our beaches. In fact, the whole of Southwest Tobago rests on ancient coral limestone deposited over hundreds of thousands of years by the tiny coral polyps that make up coral reefs.

Tobago's coral reefs are some of the best in the region, and because of its nutrient-rich coastal waters, they are also home to an impressive abundance of marine life, ranging from the microscopic to the enormous. Plankton (which often gives a green or brown tint to the surface waters during the rainy season) is the primary food for a thriving food web of marine life of all shapes and sizes. Much of it ends up as food for the massive shoals of small fry, which in turn feed large predatory fish, such as jacks, barracuda, wahoo, tarpon and tuna. The rich waters are also the reason for the massive size of some of the hard corals – such as the giant brain coral off Speyside which is over 6 m wide – and the huge barrel sponges that can be seen in the Columbus Passage south of Tobago.

Buccoo Reef is the largest coral reef in Tobago and was designated a marine park in 1973. The reef flats have wave-resistant coral species adapted to turbulent waters, such as Elkhorn Coral, and reef crests dominated by the Star Coral. In the deeper Coral Gardens the coral communities change to large colonies of brain coral, Starlet Coral and Star Coral, with many soft corals that sway in the current. Tragically, the Buccoo Reef is today a shadow of what it once was. A combination of pollution from land run-off and physical damage from reef walking and anchors has degraded much of this once majestic reef.

Source: Dr. Owen Day, Buccoo Reef Trust (http://www.mytobago.info/diving06.php) (adapted)

Responsible waste disposal in our throw-away society

Keywords
biodegradable, landfill, recycle

Did you know?
Batteries contain poisonous heavy metals such as lead, cadmium and mercury, and should not be thrown away with regular rubbish. SWMCOL collects batteries for special disposal.

Figure 15.1 *An engineered landfill site.*

Figure 15.2 *(a) Biodegradable and (b) recyclable.*

Waste in Trinidad and Tobago – from dump to managed landfill

The Solid Waste Management Company Limited (SWMCOL) started its operation in the early 1980s to deal with our rubbish in Trinidad and Tobago. Originally, rubbish was disposed of in open burning dumps which were sources of water and air pollution. These have been replaced with managed **landfill** sites.

The main landfill for Trinidad is at Claxton Bay. The site has a proper liner to reduce the risk of poisonous substances contaminating the environment. A layer of soil is put on the rubbish each day to compact it and cover it over. Any liquids that form (leachates) are collected and properly treated. The neighbouring water supplies are monitored to check for seepage contamination. When the landfill is full, it is sealed and grassed over.

But because we live on small islands, there is limited space for all the rubbish that we produce. SWMCOL needs your help in reducing the amount of rubbish sent to the landfill.

Can it be recycled?

A high-consumer society that relies on disposable items is often called a 'throw-away' society. Before putting your rubbish in the bin, ask yourself, 'Is it **biodegradable**? Can it be **recycled**?'

Biodegradable	Recyclable	Landfill
Biodegradable substances such as leftover food and garden refuse can all be turned into compost.	Many substances such as glass, paper, plastics, batteries, metal tins, etc. can be recycled.	What cannot be composted or recycled must be sent to a landfill.

Environmentalists encourage us to practice the 3 Rs:

- Reduce the amount of waste you generate by buying only what is necessary, and by buying long-lasting goods with as little packaging as possible.
- Reuse whatever you can, e.g. scrap paper, packaging, plastic bags and bottles. Try to have broken items repaired.
- Recycle cardboard, egg-boxes, plastic containers, waste paper, glass bottles and jars, etc. rather than throwing them away.

INVESTIGATION How much rubbish do my family and I produce?

Take charge of the rubbish bin at home for 1–2 weeks.

1 First, take a look at the contents of your rubbish bin(s).
   - Record how much waste your family produces on a weekly or daily basis. You can measure it by weight or bulk (volume).
   - Divide your waste into types. How much of it is recyclable and what is biodegradable?
   - Calculate how much waste each family member produces.
   - Do you think this is a lot of waste for one family? Is there room for improvement? If the answer is yes, try to reduce the amount of waste.

2 What creative ways can you think of to reduce your waste? Consider the following.
   - Are you going to try one approach (e.g. starting a compost heap), or more than one approach (e.g. starting a compost heap together with glass recycling)? Are you going to aim to reduce the volume of your rubbish or the weight of your rubbish, or both?
   - How are you going to get the members of your family to help you? Are you going to give them a pep talk about their duty to the environment, or do you need to offer rewards? Imagine if you had to pay for refuse collection by weight. What if you introduced a fining system?
   - Is there any way of earning some money from your waste?

3 After a week or more, look at whether or how well your method(s) worked.
   - It is often difficult to change your habits. How cooperative were your family members? Did anyone sabotage your efforts?
   - Were your efforts to reduce your waste successful?
   - What were the benefits and disadvantages of trying to reduce your waste?

4 Present your investigation as a poster.

Figure 15.3 *Toys like these can easily be made from recycled materials.*

Science Extra

Some plants, such as the Indian mustard plant, are able to absorb heavy metals or toxic trace elements through their roots; these plants can then be harvested and safely disposed of. The process of using plants to clean up the soil is called phytoremediation.

Figure 15.4 *An Indian mustard plant.*

Assessment

Write a one-page essay with the title 'Am I a litter bug?'. You may write as if you are asking this question of yourself or of someone else.

Oil spills and pollution

Keywords
bioremediation, crude oil

Did you know?
Although it is only the large oil spills that make the news headlines, oil spills happen daily. More than an average of 70 a day are reported in the United States.

Science Extra
Only 5% of oil that goes into the sea is spilt from tankers. The biggest input of petroleum hydrocarbons is from land sources. This includes the improper disposal of engine oil poured down drains or into the soil.

The accidental leakage of oil from damaged tankers while in transit is a regular occurrence. This map shows the sites where oil spills have taken place. In 1979, the *Atlantic Empress* collided with the *Aegean Captain* off the coast of Tobago, spilling a total of almost 90 million gallons of crude oil.

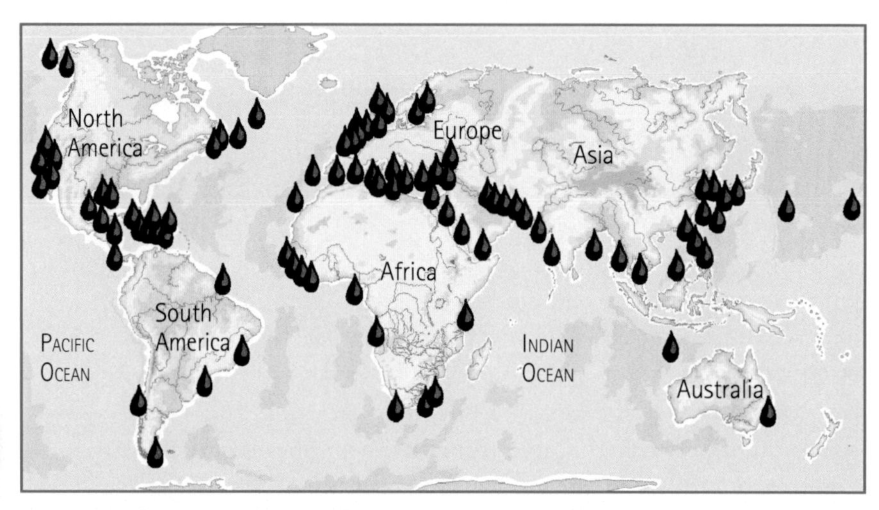

Figure 16.1 *Sites around the world where oil spills have occurred.*

The harmful effects of oil spills

Crude oil and petroleum products are complex substances that can poison fish and plankton, but the most environmental damage is caused when the oil reaches the coastline. The biggest victims of oil spills are seabirds. The sticky residue coats their feathers – this makes them lose their waterproofing and interferes with their insulation and body temperature control. Oil also poisons mammals such as manatees and seals, and can cause damage to their digestive and respiratory systems. The toll on wildlife can be very high. It is estimated that 250 000 seabirds, 28 800 sea otters, 300 seals, 250 eagles and perhaps as many as 22 orca whales were killed soon after the *Exxon Valdez* oil disaster near Alaska in 1989.

Figure 16.2 *Seabirds suffer most from oil spills. These birds have been contaminated by an oil slick from a supertanker.*

Cleaning up oil spills

The clean-up cost of oil spills is high. Because oil floats on water, the first course of action is to prevent the oil from spreading, and so floating booms are usually placed around the edges of the spill as a physical barrier. Boats are used to skim the oil off the surface, or it is sprayed with strong detergent chemicals that break the oil into small particles. The wave action of the sea helps the detergent to work. Oil deposits on the shore have to be removed using shovels, bulldozers and trucks.

The sticky tar-like deposits are difficult to remove from rocky shorelines and marshes, and so are often left to degrade slowly over time. Oil-free animals are kept away from the spill, if possible, using noise, smoke or reflectors. Oiled animals are rescued, cleaned and rehabilitated, although the capturing and rescue of mammals is often not carried out because of the stress caused to the animals.

Figure 16.3 *A team of workers cleaning up a beach polluted by an oil slick.*

Bioremediation

Bioremediation is the use of micro-organisms to degrade toxic substances into less harmful products. This method was used to clean up the shoreline after the *Exxon Valdez* oil spill in Alaska. Scientists used indigenous microbes that were able to break down the hydrocarbons in the oil into carbon dioxide and water. To supplement the microbes' diet with nitrogen and phosphorus (all organisms require carbon, nitrogen and phosphorus for growth), the scientists successfully added fertilisers to the oil.

Through bioremediation, the oil was degraded in 2–5 years. Without this intervention, scientists estimate it would have taken 5–10 years. Genetic engineering aims to improve the biodegrading activity of certain kinds of bacteria.

Operation Rescue Penguin

Following an oil spill off the coast of Cape Town, South Africa in 2000 from the ship *Treasure*, 19 500 threatened oil-free African penguins were evacuated from Dassen Island and released about 1 000 km away in Port Elizabeth. The progress of the penguins on their return journey was monitored by tagging three penguins with satellite transmitters; these penguins were nicknamed Peter, Pamela and Percy. It took the penguins about 26 days to return to their home island and by this time most of the oil slick had been cleaned up.

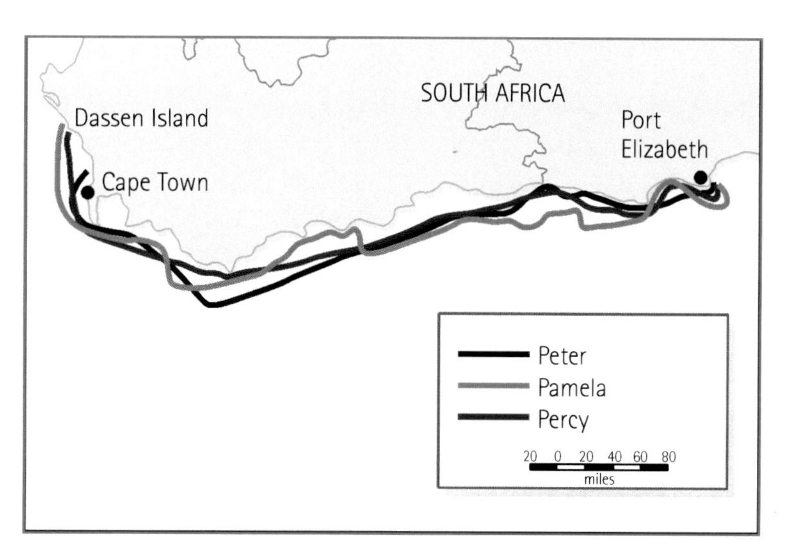

Figure 16.4 *Map of the penguin route.*

ACTIVITY

Find out more about the evacuation of penguins from Dassen Island near Cape Town to Port Elizabeth and the clean-up operation, or use the Internet to find more details on the effect of oil spills on marine life.

Figure 16.5 *A group of African penguins on a sandy shore.*

Assessment

An old oil drum has been dumped into the local pond, and is leaking oil. What damage could it do to plant and animal life, and how would you go about cleaning it up?

How environmentally unfriendly are we?

Keywords
marine ecosystem

In Unit 14 you read an extract about the damage that human activities are causing to Buccoo Reef in Tobago. Reefs consist of corals, and corals are made up of living organisms. In Trinidad and Tobago, our reefs are under attack both by nature and by human actions. Sedimentation, stemming from sediment from the large rivers of South America and Trinidad, together with sewage and other manufactured products and human activities, is having an adverse effect on our coral reef population. Legislation is in progress to try to protect and save our reefs.

Trawling is another human activity that is destroying the seabed and Trinidad and Tobago's **marine ecosystem**. Some fishermen still use nets that have been banned in Trinidad and Tobago, following worldwide standards for fishing nets. These nets catch the fish as well as other organisms not intended for catching, such as turtles, stingrays and other protected species.

Read these news clippings, which give examples of other ways in which human activities can harm the environment:

Fish stocks drastically on the wane

Sunday Express, 31 August, 1997

It may not appear that way from the heaps of kingfish and carite seen in the fish markets every week, or the fat slices of seasoned shark wrapped in Cellophane in the supermarket's frozen section – but fisheries experts and fishermen tell a different tale. Fish stocks have been depleted and some species like shark and carite are already under threat of extinction. 'It is clear that we are over-fishing our waters,' says Dr Indar Ramnarine, fisheries biologist at the University of the West Indies Department of Zoology, 'and if we continue as we are doing now, we could see the complete collapse of our commercial fisheries.'

Dead fish tell ominous chemical tale

Marine News, 10 September, 2003

The Institute of Marine Affairs (IMA) is investigating the cause of a fish kill off Point Lisas area after the discovery of large numbers of dead white mullet, moonshine, crayfish and catfish. High levels of ammonia and phosphates were measured in the water at Maracaibo.

Sewage spill threatens marine life

Nation News Barbados, 14 July, 2005

Thousands of gallons of sewage spilled into a storm drain and into the sea, causing several beaches along the South Coast to be closed as a precaution. They were reopened to the public on Tuesday. Director of the Barbados Marine Trust (BMT), Michael Webster, said that when accidents like last Friday's occurred, there was potential for damage to marine life because of bacteria and other harmful elements.

Figure 17.1 *Dead sea fish on a beach, killed by chemical poisoning.*

Figure 17.2 *Fishing with nets harms species such as turtles and stingrays.*

ACTIVITIES

1 The table below lists some of the ways that humans impact on the environment. Make a table of the examples you have covered here and in the previous units, as well as any other examples you can think of.

Human activity	Impact on environment
building roads	loss of habitat for plants and animals; noise from cars scares, wildlife away; pollution from cars kills some organisms
growing sugar cane crop	

2 Organise a class debate about attitudes towards the environment. One side should argue for exploitation, and the other side should argue for sustainability as outlined below. Consider how environmentally unfriendly we are as humans.

Pro exploitation	Pro sustainability
The environment is a resource to be used. Humans are the most important species on the planet, and we have the right to use the environment to serve our needs. Modern science and technology will always produce solutions to problems such as pollution and the shortage of resources.	We share the environment with plants and animals and it works as a system, so whatever happens to it affects us too. We have a responsibility to our children and future generations to look after our environment.

3 Look for signs of environmental degradation in your neighbourhood and identify a habitat that is suffering due to human activity. Take photos of the area to display at your school. Write a report for a local newspaper, informing the public about the environmental damage being done.

Our attitude towards the environment

As humans we can, to a certain extent, manipulate our environment for our comfort, quality of life, and convenience. But because of the fact that the human population is so large and natural resources are limited, our activities have consequences for our immediate environment and our planet as a whole. We can make choices about what we do.

Science Extra

The size of your 'ecological footprint' is the area of land that is used to support your way of living.

Assessment

Does your lifestyle gobble up the Earth's resources? Is it extravagant or modest? Complete the questionnaire 'Protecting the Earth: How big is your ecological footprint?' at www.rsc.org/education/teachers/learnnet/jesei/ecofoot/students.pdf.

18

Review topic 1

An ecosystem is a unit of living (biotic) and non-living (abiotic) parts through which energy flows and nutrients are recycled. It is made up of a particular habitat or environment such as a wetland, forest or savannah, and the community of plants and animals that live there. Complex interactions take place between the plants and animals. There is competition for resources such as food, space and light, and there are levels of feeding in the food chain and more complex food webs. Plants make their own food using the light energy of the Sun, and for this reason we call them the producers. Animals are the consumers of plants or other animals.

Oxygen is a by-product of the process of photosynthesis and is used for cellular respiration. This is the reverse process of photosynthesis, in which glucose is broken down to release energy. Everything in an ecosystem is recycled by the detrivores – the scavenging animals and the decomposers, which are the bacteria and fungi.

Because of all these complex interactions, ecosystems are in a fine balance, and can easily be disrupted. Some natural events or changes in climate have led to the extinction of plant and animal species in the past, but human activities can significantly harm the environment. Deforestation, industrial waste, the use of pesticides and fertilisers in intensive agriculture, and oil spillage into the sea all affect the survival and balance of plant and animal populations. It is important to be aware of cause and effect, and to carefully manage and protect the natural resources of our environment.

Multiple-choice questions

1 Which of the following is not an abiotic component?
   a soil
   b water
   c population
   d sunlight
2 All of the following are consumers except the
   a dove
   b snake
   c cabbage
   d worm
3 Which of the following groups of animals are predators?
   a grasshopper, snake, eagle
   b snail, grasshopper, eagle
   c lizard, grasshopper, eagle
   d eagle, lizard, snake
4 In a parasite-host relationship
   a the parasite benefits whilst the host does not
   b the host benefits whilst the parasite does not
   c neither benefits
   d both benefit
5 Different levels of a food chain are called
   a biotic components
   b trophic levels
   c populations
   d niches

Longer questions

1 a Briefly define a food chain.
   b What is the difference between a primary and a secondary consumer?
   c Which level of a food chain has the largest population and why do you think this is?
2 a Define symbiosis and give three examples.
   b Define parasitism and give two examples.
3 Use a fish, algae, an electric eel, a water beetle and a tadpole to:
   a draw a food chain and identify a primary consumer, a secondary consumer and a producer
   b draw a food web
   c explain what could happen if the water beetle was removed from the above food web.
4 Fishing trawling and the run-off or dumping of manufactured chemicals into our sea occur at an alarming rate. Discuss the consequences of both of these practices.
5 Name four endangered species in our country and suggest specific ways of protecting two of them.
6 Look at the table below and explain the relationship between forest fires in the Northern Range and floodwater levels in the Caroni Basin.

Year	Number of forest fires in the Northern Range	Average level of flood water in the Caroni Basin (m)
1991	16	1.2
1995	10	0.9
2002	4	0.5
2005	23	2.0

7 There are over half a million vehicles on the roads of Trinidad and Tobago. Each vehicle uses approximately 3 litres of oil in its engine.
   a If each driver changes oil twice a year, calculate the barrels of waste oil produced by the vehicles in Trinidad and Tobago in one year. (Note: One barrel of oil holds 150 litres.)
   b How do drivers/autoshops dispose of waste oil in Trinidad and Tobago?
   c What impact would the improper disposal of waste oil have on the environment?
8 Draw a map of Trinidad and Tobago.
   a Include forest reserves, wetlands, industries, cities/towns and dumps on the map.
   b Referring to the map, write a few recommendations to your local authorities concerning land use and environmental management.
9 'We live in a global village. Whatever we do affects our neighbours.' Discuss this statement with regard to environmental issues.
10 a List 10 environmentally friendly products/inventions.
   b Discuss two of these products/inventions.

19 Topic 2: Air and the atmosphere

The air around us

Keywords

air, atmosphere, gases, noble gases, troposphere, water vapour

Did you know?

The friction between a space shuttle and the atmosphere makes the outside surface of the shuttle heat up to thousands of degrees Celsius.

The atmosphere

Earth is surrounded by a mixture of different **gases** we call the **atmosphere**. The atmosphere is necessary for us to survive. It keeps the planet warm, provides us with oxygen to breathe, and protects us from harmful radiation from the Sun.

Scientists divide the atmosphere into different layers, giving each layer a different name.

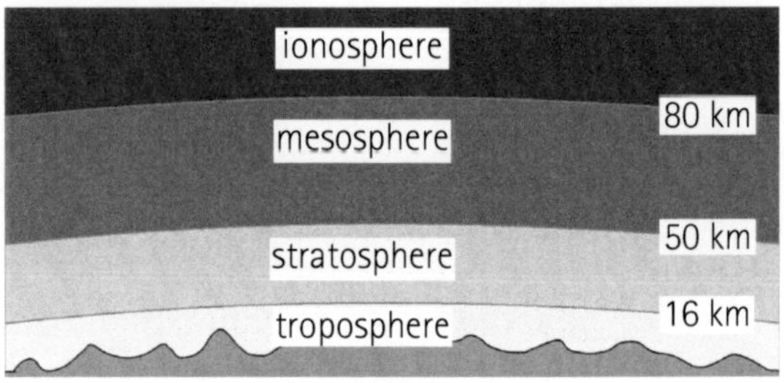

ionosphere
80 km
mesosphere
50 km
stratosphere
16 km
troposphere

Figure 19.1 *The different layers of air in the atmosphere.*

Although we don't feel it, the weight of all that air presses down on us, and we call this phenomenon air pressure. You will study air pressure in Unit 76. The atmosphere reaches almost 600 km beyond Earth. As we move farther away from Earth, the layers become less and less dense, and from about 100 km above Earth, we reach the part we call space.

The part of the atmosphere that we are most interested in is the part nearest to Earth – the **troposphere**. This layer is where the weather takes place, and it is approximately 16 km thick.

Gas	Percentage (%)
Nitrogen	78%
Oxygen	21%
Carbon dioxide	0.03%
Noble gases	Approx 1%

Air is a mixture of gases

Air is a mixture of different gases, mostly nitrogen and oxygen. It also contains carbon dioxide and some unusual gases we call the **noble gases**. The table on the left gives you the percentages of these gases in the air.

Science Extra

The mixture of gases in air can be separated by the fractional distillation of liquid air. The air is cooled until it forms a liquid at around −200 °C, and then it is slowly heated. Each part, or fraction of the mixture, boils off at a different temperature. Nitrogen boils off at −196 °C and oxygen at −183 °C.

ACTIVITY

1 Draw a pie chart to show the composition of air using the values in the table in the margin. (Use a protractor if you have one. Otherwise, estimate the sizes of the segments of the pie.)
2 Complete the following:
  a Which gas is the most essential for animal life? Is it the most abundant element in the air?
  b Almost four-fifths of the air is made up of …
  c Although plants depend on … for photosynthesis, it makes up less than 0.1% of the air.

INVESTIGATION How much oxygen is there in air?

This investigation can be written up for assessment (see below).

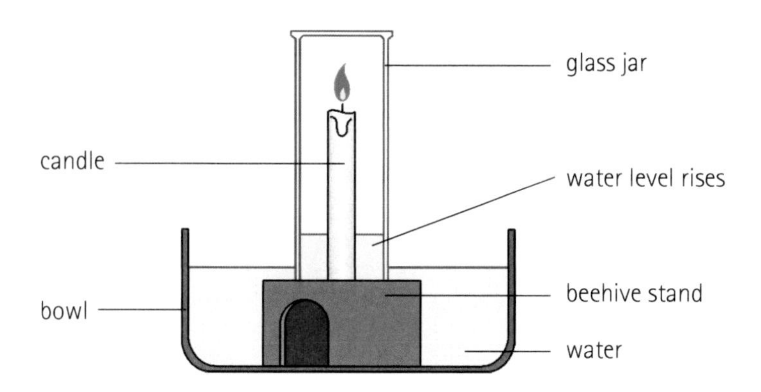

Figure 19.2 *Apparatus for investigating the amount of oxygen in the air.*

1 Place a candle on the beehive stand in a bowl of water.
2 Light the candle and place a glass jar over the candle.
3 The oxygen is used up as the candle burns and the water rises up to fill the space where the oxygen was. Knowing the percentage of oxygen in the air, can you predict roughly how high the water will rise in the jar?
4 Measure how far the water rises and the height of the jar. Use these measurements to calculate the percentage of the way up the jar the water rose. This corresponds to the percentage of oxygen in air.
5 How accurate is your result and what do you think affects the accuracy of the experiment the most?

Safety: Always take care not to start a fire when working with lighted matches and candles. Always clear your desk of papers and books before you start.

Air also contains water vapour

Water is also present in the air in the gaseous form of **water vapour**. The reason that no percentage is given for water vapour is because the amount can vary, depending on how dry or humid the air is.

In Unit 26 you will look at factors that influence the amount of moisture in the air, and the ways you can show the presence of water vapour.

Assessment

Write up the Investigation as a report, using *Aim of experiment*, *Method*, *Results* and *Conclusion* as headings.

Figure 19.3 *Water vapour in the air is usually invisible, but clouds (above) or mist (below) are visible forms.*

20

Oxygen

Keywords

atom, combustion, corrosion, molecule, oxide, oxygen, respiration

Did you know?

The photosynthesis of early bacteria and plants produced the oxygen that is now in our atmosphere. Organisms on Earth went through millions of years of evolution before they could make use of it.

Oxygen as a molecule

Oxygen consists of two oxygen **atoms** joined together, and is therefore written in symbol form as O_2. When atoms are joined as a unit, the particle is called a **molecule**.

Figure 20.1 *A model of an oxygen molecule.*

A test for oxygen

A glowing coal or a glowing splint will relight in the presence of oxygen. Try out some other combustion reactions in the investigation on page 45.

Oxygen is the reactive part of air

Oxygen is the most reactive of the gases in the air. This means that it combines easily with other substances through chemical reactions. It is because of oxygen in the air that iron rusts and fuels are able to burn. Processes such as **respiration**, **corrosion** (rusting) and **combustion** are all examples of reactions with oxygen, and can be thought of as forms of 'burning'.

Figure 20.2 *Respiration releases energy.*

Respiration

The oxygen that we breathe from the air is necessary for the process of respiration (which you learnt about in Unit 8). Glucose (made from sugars in food) reacts with oxygen dissolved in the blood to release energy. The products of respiration are carbon dioxide and water:

glucose + oxygen → carbon dioxide + water + energy

Figure 20.3 *Oxygen combines with metal during corrosion.*

Corrosion (rusting)

Many metals corrode. They react slowly with the oxygen in the air to form **oxides**, and the surface is 'eaten' away, e.g.:

iron + oxygen → iron oxide

$4Fe + 3O_2 \rightarrow 2Fe_2O_3$

The presence of water in the air speeds up the corrosion process.

Figure 20.4 *Oxygen combines with fuel during combustion.*

Combustion

Fuels, such as wood and methane gas, are carbon-containing substances that burn easily and release lots of heat energy. They combine with oxygen in the combustion process to produce carbon dioxide and water, e.g.:

methane + oxygen → carbon dioxide + water + heat

$CH_4 + 2O_2 \rightarrow CO_2 + 2H_2O$ + heat

Many elements, such as magnesium and sulphur, combine with oxygen to form oxides.

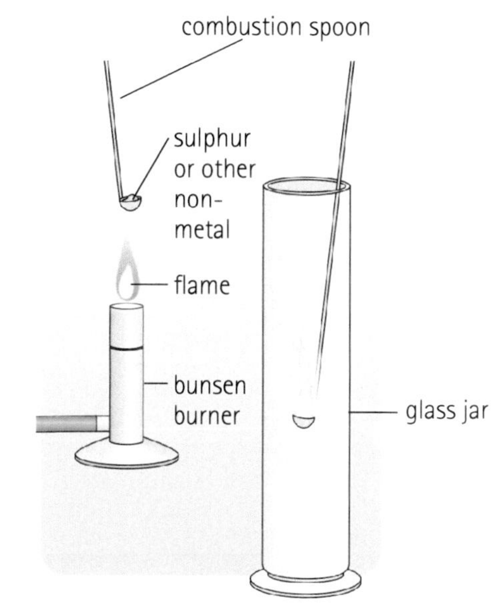

Figure 20.5 *Investigating combustion.*

INVESTIGATION The burning of elements in oxygen

1. Your teacher will first demonstrate the combustion of a piece of magnesium. Magnesium is a metal used in fireworks that burn with a bright white flame. Observe how the magnesium changes to a grey, ash-like substance – magnesium oxide.
2. Then place a small amount of sulphur on the end of a combustion spoon. Sulphur is a non-metal. Make a note of the colour and the smell.
3. Carefully heat the sulphur in the flame of a bunsen burner. As soon as the sulphur catches fire, place it inside a glass jar. Make a note of the colour of the flame. Can you see the sulphur dioxide gas produced?
4. Clean the spoon, and repeat the experiment with another non-metal such as phosphorus, or metals in powdered form, such as copper, zinc, calcium or iron. Again, make a note of what the substances look like before and after burning, and note the colour of the flames as they burn.
5. Draw and fill in a table of your results as shown below:

Element	Colour of flame	Oxide formed
phosphorus (non-metal)	yellow flame	phosphorus oxide (white smoke)

Safety: The combustion spoon becomes hot when you place it in the bunsen flame. Make sure that you wear gloves, safety goggles, and use a heat-proof mat under the bunsen burner.

Science Extra

The higher the altitude, the thinner the atmosphere becomes, and so the amount of oxygen in the air decreases. Mountaineers must climb high peaks slowly to allow their bodies time to adapt. If they climb too fast they risk altitude sickness and even death.

Uses of oxygen

- Oxygen can be extracted from air and used in its pure form in industry, to convert pig iron (a kind of iron that is not pure) into steel.
- Oxygen, mixed with acetylene, produces a very hot flame, and can be used for welding as shown in Figure 20.6.
- In hospitals, oxygen helps patients who have breathing difficulties.
- It is also used to restore life to polluted rivers.

Figure 20.6 *Oxyacetylene burns with a flame that is hot enough to cut through metal.*

Assessment

Research the manufacture of liquid oxygen at Point Lisas industrial estate.

21

Carbon dioxide

Keywords

carbon, carbon dioxide, compound, element

Did you know?

When carbon is heated with carbon dioxide (CO_2), carbon monoxide (CO) is produced. Carbon monoxide is a very poisonous gas and is often present in car exhaust fumes. It is therefore very dangerous to leave a car engine running in an enclosed space such as a garage.

Carbon dioxide is also used in blast furnaces to extract iron from iron ore (iron oxide) by removing the oxygen i.e.:

iron ore + carbon monoxide → iron + carbon dioxide

$Fe_2O_3 + 3CO \rightarrow 2Fe + 3CO_2$

Compare this with the equation that shows the rusting of iron on page 44. (Remember that iron ore is iron in a naturally rusted state.)

Carbon dioxide is a compound

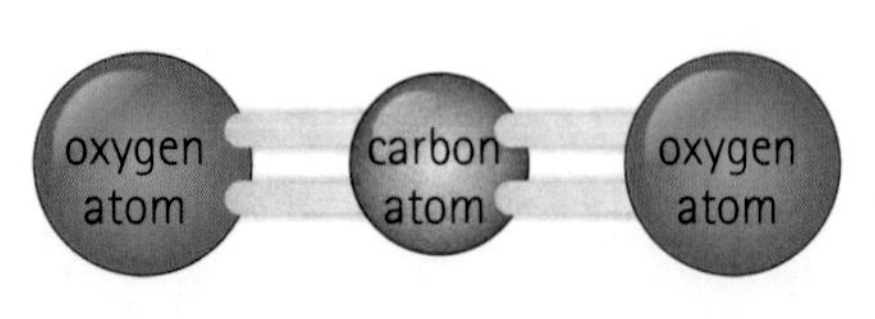

Figure 21.1 *A model of a carbon dioxide molecule.*

Whereas oxygen (O_2) is described as an **element** because it is made up of molecules of the same atoms, **carbon dioxide** is described as a **compound** because it is made up of two different types of atoms. A molecule of carbon dioxide consists of one atom of **carbon** joined to two atoms of oxygen (CO_2).

A test for carbon dioxide

Carbon dioxide turns limewater from clear to milky, and therefore limewater is often used to test for the presence of this gas. In Unit 20 you learnt that carbon dioxide is a by-product of respiration and the combustion of fuels. Now you can test for it yourself.

Testing for carbon dioxide as a product of respiration

Blow bubbles through a straw into a glass container filled with limewater and observe what happens.

INVESTIGATION **Testing for carbon dioxide as a product of burning a carbon substance**

All substances that contain carbon burn with oxygen to give carbon dioxide and water. Here you will burn a piece of wood and test for both carbon dioxide and water. Blue cobalt chloride paper turns pink in the presence of water.

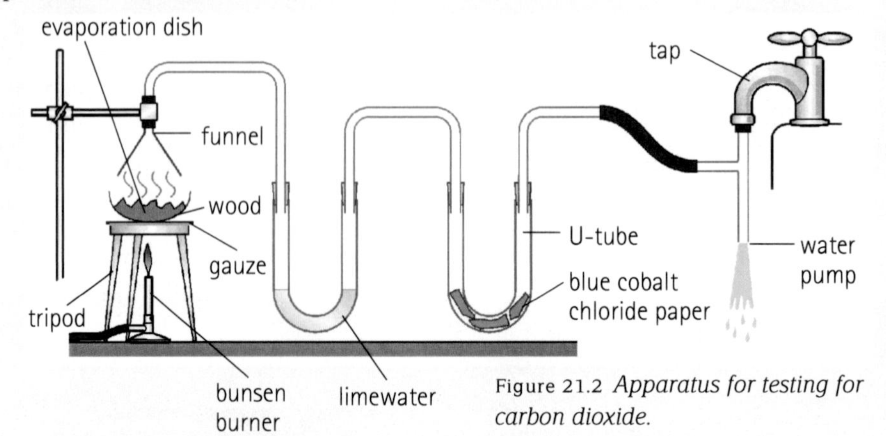

Figure 21.2 *Apparatus for testing for carbon dioxide.*

1 Set up your apparatus as in Figure 21.2.
2 Start the water suction pump. This ensures that any products from the combustion are collected.
3 Light the piece of wood and observe what happens to the limewater and the blue cobalt chloride paper.

Safety: Always take care when burning things. Wear safety goggles and use a heat-proof mat.

Science Extra

Micro-organisms, such as yeast, use the process of fermentation to produce energy from sugars. Unlike the process of respiration, oxygen is not needed.

Uses of carbon dioxide

Carbon dioxide is a useful gas. It is used in fizzy (carbonated) drinks. In some drinks, such as beer or naturally made ginger beer, the carbon dioxide is produced by yeasts, in a fermentation process. In other drinks, the carbon dioxide gas is dissolved in the drink under pressure. When the container holding the drink is opened, the pressure decreases, and the gas can escape from the liquid, causing it to fizz as shown in Figure 21.4.

Carbon dioxide is also stored under pressure in cylinders and can then be used as a fire extinguisher. There is a colour-coding system for fire extinguishers, and cylinders containing carbon dioxide are usually coloured red with a black band above the operating instructions.

Pure carbon dioxide gas is heavier than air (remember that air is a mixture of gases) and it forms a 'blanket' over the flames, displacing the air (which has oxygen in it) from the source of the fire. Unlike water, or chemical fire extinguishers, carbon dioxide is a non-messy type of extinguisher. But take note: remember how vigorously you saw magnesium burn in oxygen? Combustible metals such as magnesium and sodium can also burn fiercely in the presence of nitrogen or carbon dioxide. For these fires, sand or metal extinguishers must be used.

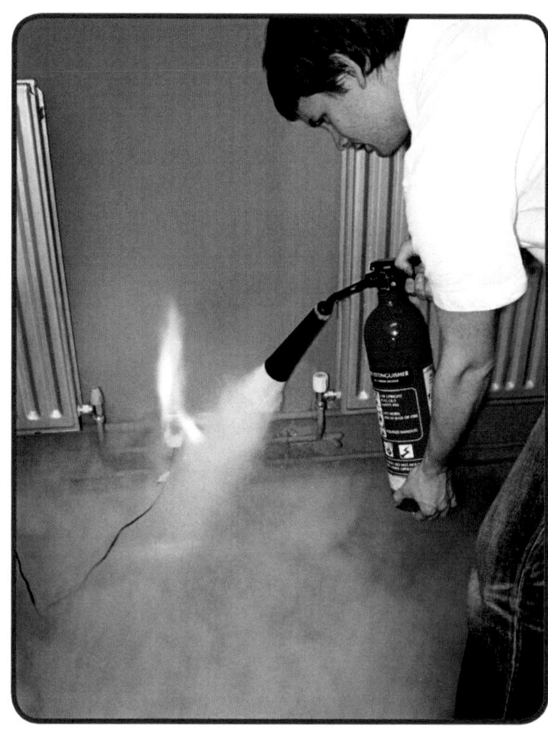

Figure 21.3 *A carbon dioxide fire extinguisher puts out an electrical fire.*

Assessment

Research some other uses of carbon dioxide. Do you know of a use (and name) for carbon dioxide, not as a gas, but in its solid form? Find out about ways (natural or industrial) in which carbon dioxide can be produced.

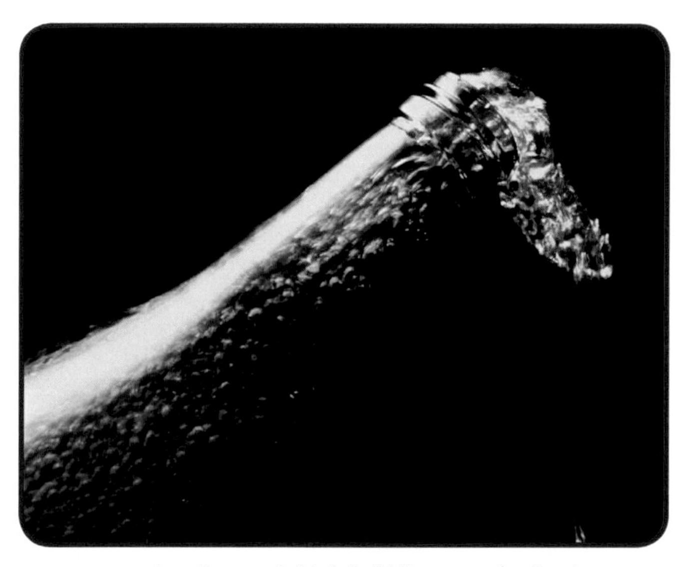

Figure 21.4 *A carbonated drink bubbling out of a bottle.*

22 Carbon dioxide and the carbon cycle

Keywords
carbon cycle, DNA, protein, RNA

Did you know?
The biggest source of carbon dioxide on Earth is in carbonate rocks.

Although carbon dioxide makes up only 0.03% of the atmosphere, plant and animal life depend on it. The **carbon cycle** explains how carbon dioxide is produced, and it also explains how carbon gets recycled through our environment.

The carbon cycle

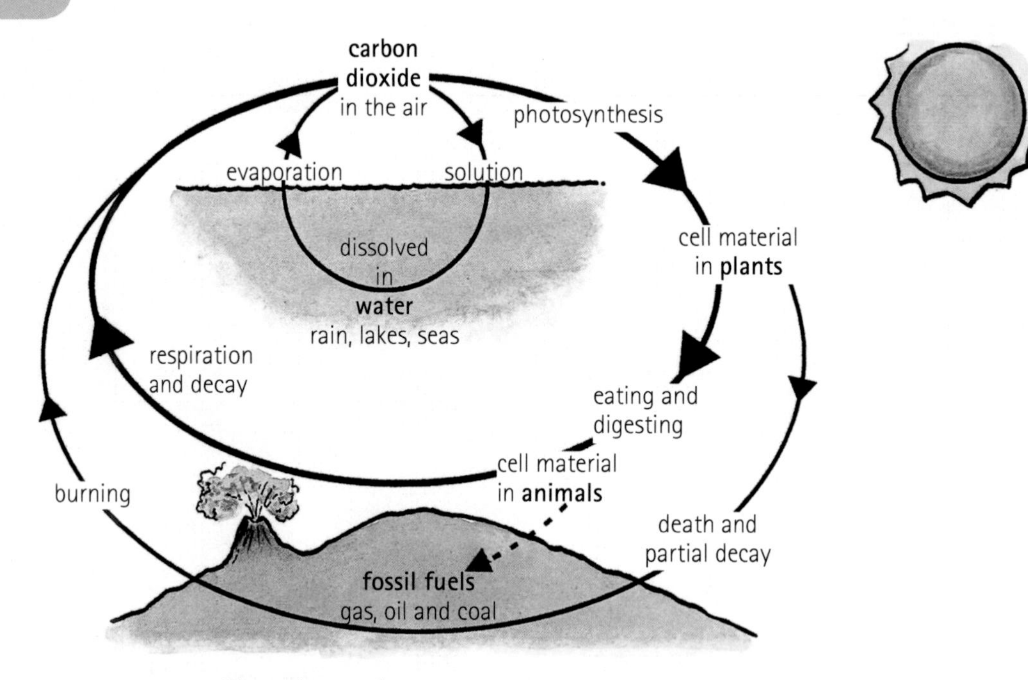

Figure 22.1 *The carbon cycle.*

Carbon dioxide is removed from the atmosphere by:

- photosynthesis
- carbon dioxide gas dissolving in water.

Carbon is released into the atmosphere in the following ways:

- respiration by plants and animals
- plant and animal decay
- combustion or burning
- dissolved carbon dioxide evaporating in warm sea water
- volcanic eruptions.

ACTIVITIES

1 Carefully read about the ways that carbon dioxide is released into and removed from the atmosphere. Identify each of these steps on the diagram of the carbon cycle.
2 Working in small groups, prepare a collage to illustrate the carbon cycle. Use pictures from newspapers and magazines.

All living things are carbon based

Carbon is an important element in nature because all living things are made mainly of carbon. Carbon atoms are joined together with atoms of other elements such as oxygen, hydrogen, nitrogen and phosphorus to make up large complex molecules such as **DNA** (deoxyribonucleic acid), **RNA** (ribonucleic acid), **proteins** and carbohydrates.

Science Extra

Carbon can form so many different compounds by combining with other elements because each carbon atom has four electrons in its outer shell. These electrons are free to pair with the outermost electrons of other carbon atoms, or of atoms of elements such as hydrogen, oxygen and nitrogen.

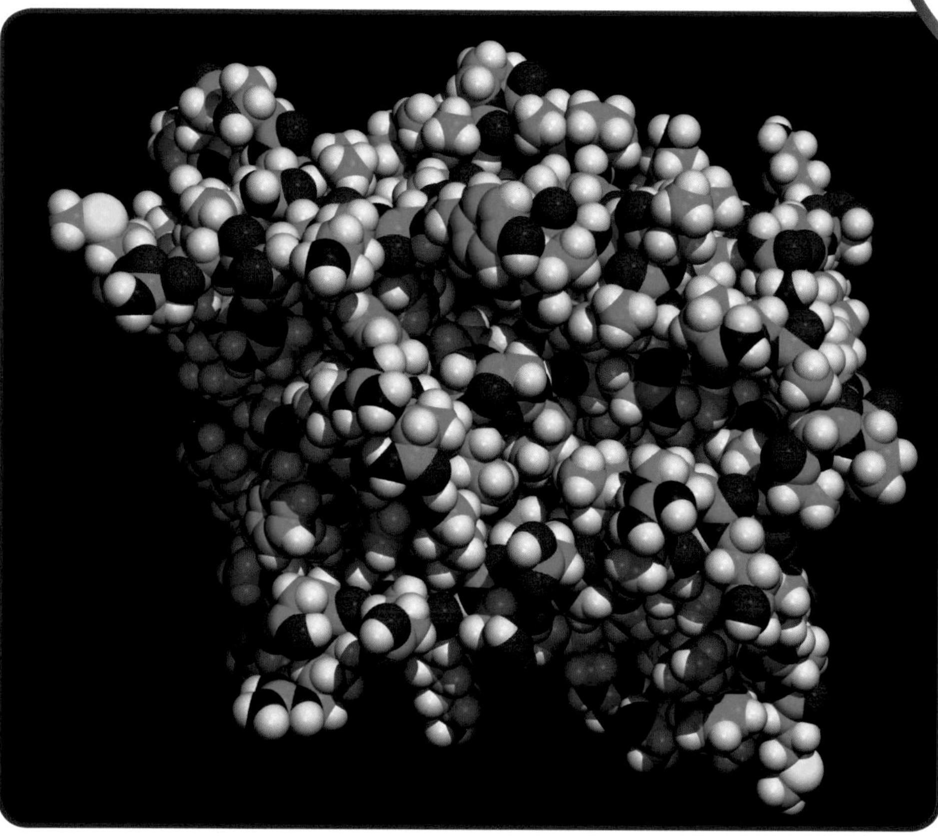

Figure 22.2 *Protein molecules are one of the building blocks of life. The carbon atoms are shown in green in this model of the human growth hormone, somatotrophin. Hormones are protein molecules.*

Assessment

Our bodies are 70% water and 18.5% carbon. Minus the water, carbon makes up more than 60% of a body's dry weight. Name three types of complex carbon-based molecules that the human body is made of.

23

Nitrogen

Keywords

chemical bond, cryopreservation, nitrogen

Did you know?

High levels of nitrogen in scuba divers' blood can have a similar effect to alcohol. The phenomenon is called nitrogen narcosis.

Like oxygen, **nitrogen** gas occurs in the air in element form, as N_2 molecules. Unlike oxygen, nitrogen gas is relatively unreactive. The diagram of the nitrogen molecule shows why. You can see that the atoms of the N_2 molecule are joined together by three **chemical bonds** (called a triple bond). This makes the bonds of the N_2 molecule very strong and the N_2 molecule itself unreactive. Nitrogen gas can be useful because of its unreactivity.

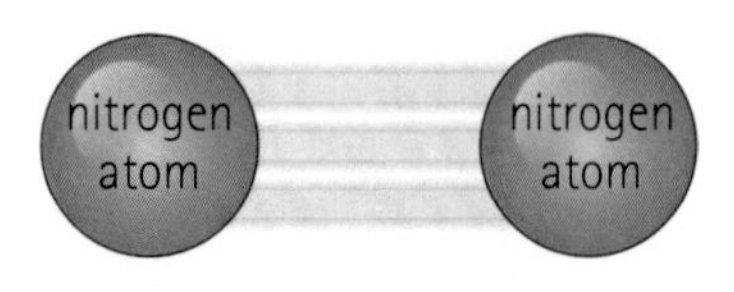

Figure 23.1 *A model of a nitrogen molecule with its triple bond.*

Uses of nitrogen

In its liquid form, nitrogen is very cold (colder than –196 °C), and is used for 'snap-freezing' to ensure the freshness of vegetables such as peas.

In the manufacturing industry, liquid nitrogen is used for 'shrink-fitting'. This is when a metal part is contracted by freezing, and then fitted inside another part. It then expands as it warms, to make a tight fit.

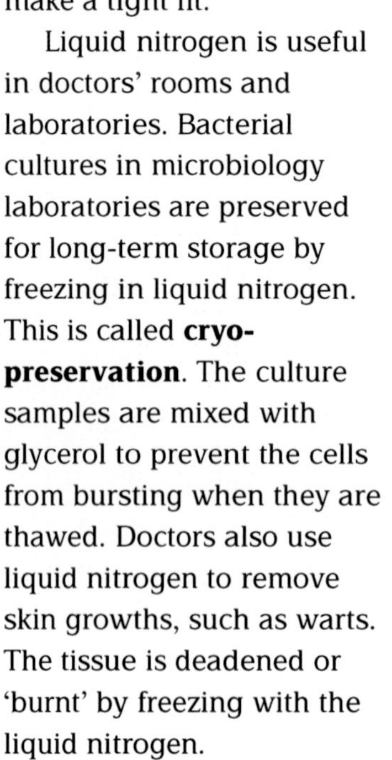

Liquid nitrogen is useful in doctors' rooms and laboratories. Bacterial cultures in microbiology laboratories are preserved for long-term storage by freezing in liquid nitrogen. This is called **cryo-preservation**. The culture samples are mixed with glycerol to prevent the cells from bursting when they are thawed. Doctors also use liquid nitrogen to remove skin growths, such as warts. The tissue is deadened or 'burnt' by freezing with the liquid nitrogen.

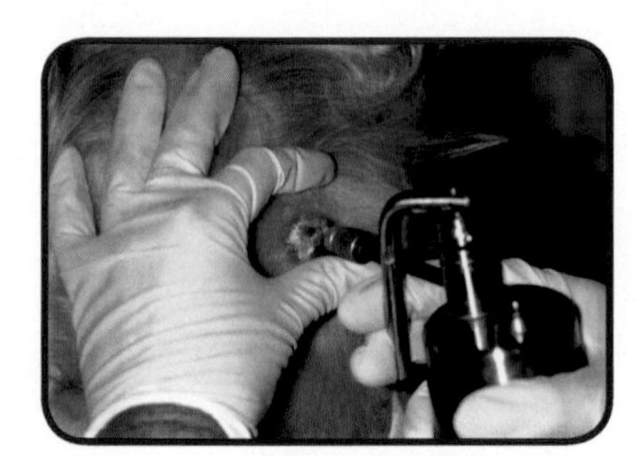

Figure 23.2 *A surgeon using liquid nitrogen to remove a wart from a woman's temple.*

Figure 23.3 *A plant physiologist about to freeze a sample of seeds in liquid nitrogen.*

Nitrogen for making ammonia

Nitrogen is used to make ammonia gas through this reaction, which requires a high temperature and high pressure:

nitrogen + hydrogen ⇌ ammonia

$$N_2(g) + 3H_2(g) \rightleftharpoons 2NH_3(g)$$

The symbol (g) indicates that the reactants and products of the chemical reaction are in the gaseous state. As you will see in the next unit, ammonia is an important ingredient in the manufacture of nitrate fertilisers such as ammonium sulphate (NH_4SO_4).

Science Extra

The production of ammonia from hydrogen and atmospheric nitrogen is called the Haber process, named after the German chemist Fritz Haber. The patent rights for this process, which Haber sold to a chemical company in the early 1900s, made him a millionaire.

How can nitrogen in the air be harmful to divers?

When divers are deep under water (and therefore at high pressure), the nitrogen that they breathe from compressed air cylinders dissolves in their blood. If they surface too quickly, without adjusting to the pressure changes, the nitrogen gas forms bubbles in the blood, blocking the blood vessels, and causing pain. This is called decompression sickness or 'the bends', and it can be dangerous.

To help you understand this, do the following activity. Think of the contents in a bottle of cooldrink as the scuba diver's blood, and of the carbon dioxide bubbles as nitrogen bubbles.

Figure 23.4 *A scuba diver under water with a compressed air cylinder for breathing.*

ACTIVITY

1. Take a close look at an unopened bottle of carbonated cooldrink. Do you see gas bubbles in the liquid?
2. Now open the bottle. What do you observe?
3. Is it the opened or unopened bottle of cooldrink that represents the diver's blood deep under water? Where does the nitrogen that dissolves in the diver's blood come from?

The pressure inside the unopened bottle is greater than the pressure inside the opened bottle. The greater the pressure applied to a liquid, the more easily gas dissolves in it. Another way of putting this is: the solubility of gases in water increases as pressure increases. (You will learn about pressure in air and liquids in Topic 5.) Once opened, the pressure is lower in the bottle, and the carbon dioxide gas comes out of solution, escaping as bubbles.

Assessment

Write a few sentences on the most interesting or important fact you learnt about nitrogen in this unit.

24 Nitrogen and the nitrogen cycle

Keywords
denitrification, nitrogen cycle, nitrogen fixation, rhizobia

Did you know?
Trinidad and Tobago has industrial plants at the Point Lisas estate that produce urea and ammonia. Our country is one of the largest producers of fertilisers in the world.

After carbon, nitrogen is the second most common element in living things. Complex molecules such as DNA and the proteins that make up the building blocks of life contain nitrogen. Look at Figure 22.2 on page 49. The nitrogen atoms are shown in blue in the model.

The nitrogen cycle

Although nitrogen makes up 78% of air, and it is a common element in living things, the natural processes by which nitrogen is extracted from the air are quite complicated. Remember that the triple bond of the N_2 molecule makes nitrogen unreactive, and therefore difficult to use. The **nitrogen cycle**, shown in Figure 24.1, explains how nitrogen is circulated between the atmosphere, Earth, and living things (plants and animals). By the process of **nitrogen fixation**, nitrogen gas is turned into ammonia and nitrates. By the process of **denitrification**, nitrates are turned back into nitrogen gas.

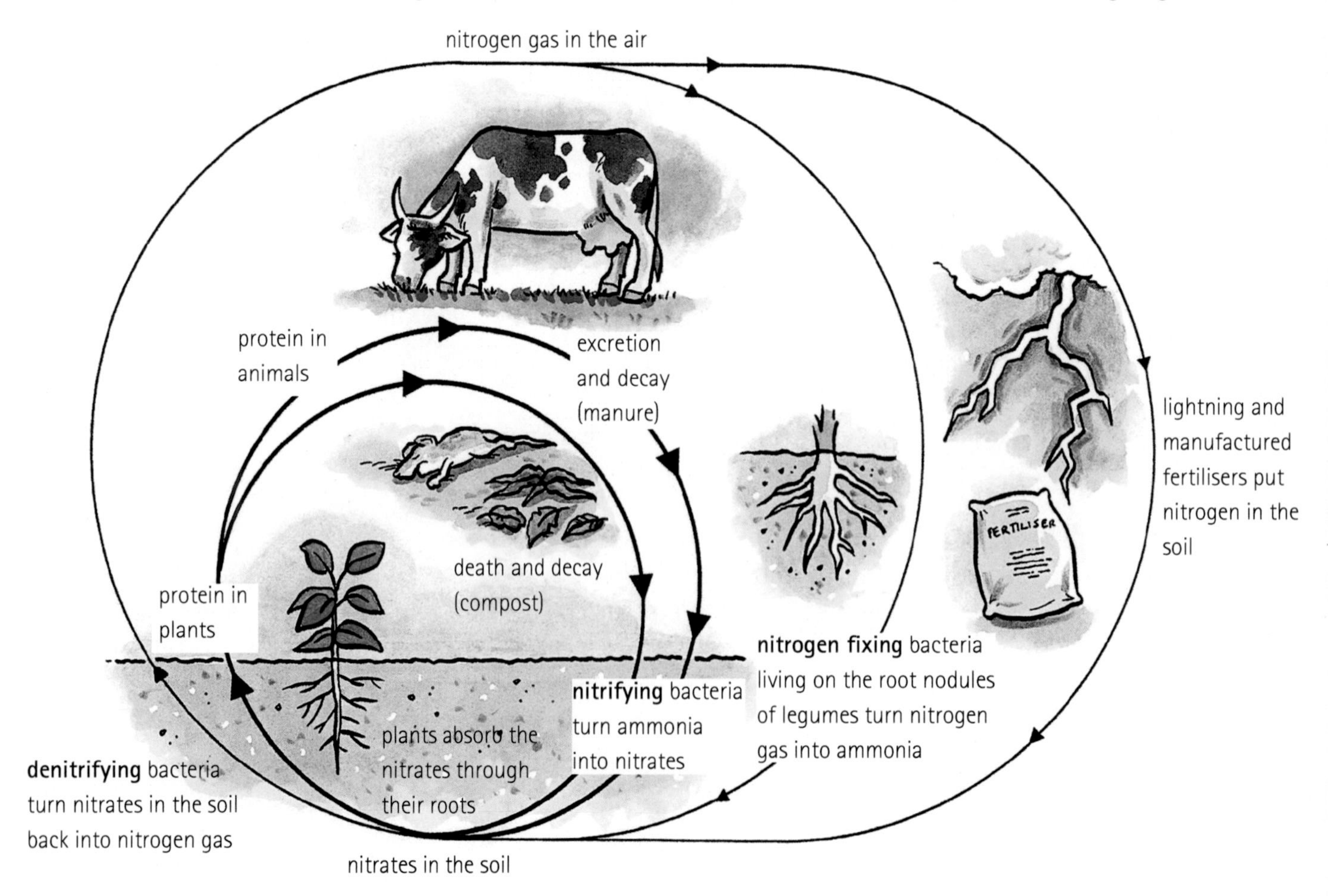

Figure 24.1 *The nitrogen cycle.*

Putting the nitrogen back into the soil – compost, manure and fertiliser

Although nitrogen is plentiful in the atmosphere, there is often a shortage in the soil, and this can limit plant growth. Also, when crops are harvested, the nitrogen that the plants have taken out of the soil is not recycled back into the soil as it usually is when plants decompose naturally, and it must therefore be replaced. Farmers can get nitrogen back into the soil by adding nitrates, natural manure or compost, or artificial fertiliser to it. Or they can grow leguminous crops, which they plough back into the soil.

Legumes, rhizobia and root nodules

Legumes are plants such as peas, soybeans and clover that bear their seeds in pods. Nitrogen-fixing bacteria called **rhizobia** live on the roots of these plants, and convert nitrogen from the air into ammonia. They provide the plant with a usable form of nitrogen and in return get their nutrients from the plant. This form of dependence, which benefits both the bacteria and the plant, is known as symbiosis. The nodules that form on the roots of the leguminous plants are abnormal growths that are caused by the bacteria themselves as a result of 'infection'.

Science Extra

Almost all the world's ammonia is made by the Haber process, which takes nitrogen from the air and combines it directly with hydrogen. The ammonia is used to make nitric acid, and the nitric acid is then processed into nitrate salts, which are used to make fertilisers.

ACTIVITIES

1 This table shows the yield of a crop in successive years when the nitrogen in the soil is not replenished.

Yield of corn (kg/Ha)	Time (year)
2 000	2001
1 200	2002
900	2003
410	2004

a What conclusions can you draw about the effect of nitrogen depletion on crop yields?

b List the ways in which nitrogen can be put into the soil. Which of these methods are organic? Consider the advantages and disadvantages of each.

2 Your teacher will show you a leguminous plant. Make a diagram, showing the root nodules, and answer the following questions.

a How would you define a leguminous plant?

b What causes the formation of the nodules?

c What is the name of this natural process in which nitrogen is turned into ammonia?

d Where does the nitrogen come from?

Figure 24.2 *Root nodules on the roots of a pea plant allow the plant to use free nitrogen in the atmosphere and soil.*

Assessment

Close the textbook and try to draw the nitrogen cycle from memory. If you get stuck, see if you can work it out. Check your drawing afterwards.

The noble gases

Keywords

argon, atomic number, atomic weight, electron, electron shell, helium, inert, neon, neutron, noble gas, periodic table, proton, valency

Did you know?

Neon was discovered by two British scientists in 1898. This gas takes its name from the Greek word *neos* meaning *new*.

Which gases are noble?

The last 1% of air is made up of a group of unreactive gases called the **noble gases**. There are six noble gases, and the percentage of each of them is shown alongside. **Helium**, **neon** and **argon** are the most common.

Noble gas	% volume in air
Helium	5.24×10^{-4}
Neon	1.18×10^{-3}
Argon	9.34×10^{-1}
Krypton	1.14×10^{-4}
Xenon	8.70×10^{-6}
Radon	6×10^{-18}

Table 25.1 *The relative percentage of noble gases in the air.*

Why are the noble gases unreactive?

Look at the **periodic table** and identify the noble gases. All the noble gases have the maximum number of **electrons** possible in their outermost shell (two for helium, eight for all the others). Remember that atoms have a central, positively charged nucleus containing particles called **protons** and **neutrons**. The nucleus is surrounded by layers or orbits of negatively charged electrons. These orbits are also called **electron shells**. The number of electrons in an atom's outermost shell is what is called the valence number or **valency,** and this determines how reactive it is. The noble gases have a full outer shell of electrons, which makes them unreactive or **inert**.

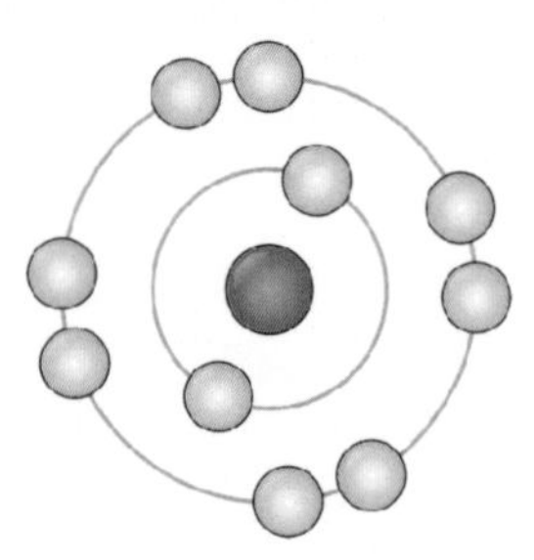

Figure 25.2 *A diagram illustrating valency. The first orbit is full when it contains two electrons. The second orbit is full when it contains eight electrons.*

KEY: 29 / **Cu** / 63,5 — Atomic number / Symbol / Relative atomic mass (approximately); shaded = noble gases

I	II											III	IV	V	VI	VII	0
1 **H** 1																	2 **He** 4
3 **Li** 7	4 **Be** 9											5 **B** 11	6 **C** 12	7 **N** 14	8 **O** 16	9 **F** 19	10 **Ne** 20
11 **Na** 23	12 **Mg** 24											13 **Al** 27	14 **Si** 28	15 **P** 31	16 **S** 32	17 **Cl** 35,5	18 **Ar** 40
19 **K** 39	20 **Ca** 40	21 **Sc** 45	22 **Ti** 48	23 **V** 51	24 **Cr** 52	25 **Mn** 55	26 **Fe** 56	27 **Co** 59	28 **Ni** 59	29 **Cu** 63,5	30 **Zn** 65	31 **Ga** 70	32 **Ge** 73	33 **As** 75	34 **Se** 79	35 **Br** 80	36 **Kr** 84
37 **Rb** 86	38 **Sr** 88	39 **Y** 89	40 **Zr** 91	41 **Nb** 92	42 **Mo** 96	43 **Tc**	44 **Ru** 101	45 **Rh** 103	46 **Pd** 106	47 **Ag** 108	48 **Cd** 112	49 **In** 115	50 **Sn** 119	51 **Sb** 122	52 **Te** 128	53 **I** 127	54 **Xe** 131
55 **Cs** 133	56 **Ba** 137	57 **La** 139	72 **Hf** 179	73 **Ta** 181	74 **W** 184	75 **Re** 186	76 **Os** 190	77 **Ir** 192	78 **Pt** 195	79 **Au** 197	80 **Hg** 201	81 **Tl** 204	82 **Pb** 207	83 **Bi** 209	84 **Po**	85 **At**	86 **Rn**
87 **Fr**	88 **Ra** 226	89 **Ac** 227	104 **Rf**	105 **Db**	106 **Sg**	107 **Bh**	108 **Hs**	109 **Mt**	110 **Ds**	111 **Rg**	112 **Cp**						

58 **Ce** 140	59 **Pr** 141	60 **Nd** 144	61 **Pm**	62 **Sm** 150	63 **Eu** 152	64 **Gd** 157	65 **Tb** 159	66 **Dy** 163	67 **Ho** 165	68 **Er** 167	69 **Tm** 169	70 **Yb** 173	71 **Lu** 175
90 **Th** 232	91 **Pa**	92 **U** 238	93 **Np**	94 **Pu**	95 **Am**	96 **Cm**	97 **Bk**	98 **Cf**	99 **Es**	100 **Fm**	101 **Md**	102 **No**	103 **Lr**

Figure 25.3 *The periodic table.*

ACTIVITY

Use the information on the previous page to answer the following questions.

1 Which noble gas is the most abundant in the air, and which is the least?
2 Which group in the periodic table do the noble gases belong to?
3 How many electrons does helium have in its outer shell?
4 What is the number of electrons in the outer shell of neon, argon, krypton, xenon and radon?
5 Is the number of electrons in the outer shell of the different groups of elements the same?
6 Name the term for the number of electrons in the outer shell.
7 Which of the noble gases has the lowest atomic number (i.e. number of protons in the nucleus) and which has the highest atomic number?

Figure 25.4 *Light bulbs are filled with argon. The filament in a bulb is heated by electricity. As it gets hot it gives off light, and because the argon is unreactive, the filament does not burn.*

What are the noble gases used for?

As the noble gases are almost completely unreactive, they can be useful and safe to use.

Helium balloons are used by the weather service to carry equipment high up into the atmosphere. Balloons filled with helium float because pure helium gas is lighter, or less dense, than air. This is because helium has a lower atomic weight than nitrogen, oxygen and carbon dioxide. Helium airships were used as a form of air transport in the early part of the 20th century between Britain and America.

Look at the periodic table and compare the atomic number of helium with the other elements that make up air. There is a relationship between the **atomic number** (the number of protons in an atom) and the **atomic weight** (the average mass of the atom of a particular element). The greater the atomic number of the element, the greater its atomic weight. Knowing this, do you think a radon-filled balloon would float?

Hydrogen gas is also very light (have a look at its atomic number), but it is reactive, and so it is not suitable for filling balloons because of the risk of explosions.

Figure 25.5 *Neon lights are often used in advertising. A neon light is filled with neon gas. An electrical charge is passed through the gas, making it glow. The tubes can be made into all sorts of shapes.*

Assessment

Do research on one or more of the noble gases. Find out, for example, how their names were derived (you can find this out in a good dictionary), or what colour each of the noble gases would make if electricity was passed through it. Present your findings as a poster.

Figure 25.6 *A passenger-carrying helium-filled airship.*

26

Water vapour in the air

Keywords

capillary action, humidity, hygrometer, hygroscopic, water vapour

Did you know?

Lightning will not occur if there is no water vapour in the air.

Have you noticed how table salt becomes damp during the rainy season? Or how a packet of opened biscuits can lose their crispness? This is because of **water vapour** in the air. The amount of moisture or water vapour in the air is also called **humidity**.

Humidity

Humidity varies and usually depends on the temperature. The warmer the air, the more water vapour it can hold. Generally, humidity is high where the weather is warm and there is a source of water for evaporation, e.g. at places near to the sea, or during the rainy season. As the air cools, the water vapour condenses (liquefies), and the amount of water vapour in the air is reduced. There is a limit to how much water vapour the air can hold before it condenses, and this is the principle behind cloud formation and rain.

A **hygrometer** is an instrument used to measure humidity. Make a simple hygrometer in the following activity.

ACTIVITY

1 Follow the diagram to set up a 'wet and dry bulb' hygrometer. **Capillary action** makes the water move up the string, i.e. the string works as a wick. Remember that evaporation causes cooling, so the thermometer bulb with the damp cloth wrapped around it (wet bulb) will record a lower temperature than the other (dry bulb). The lower the humidity, the greater the rate of evaporation.

2 Record the difference in temperature between the two thermometers. Complete the following: The bigger the difference in temperature, the ... (lower/higher) the humidity.

3 Repeat the measurement at another time. Do you predict the humidity to be more or less than before? Give a reason for your answer and compare this with your result.

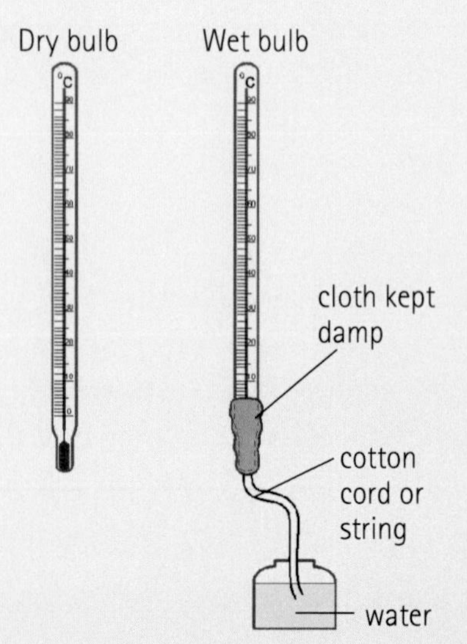

Figure 26.1 *A home-made hygrometer.*

Keeping the air dry with silica gel

Science Extra

The more water vapour there is in the air, the lower the air pressure. This is because water vapour molecules are lighter than the mixture of gases that make up air. The molecular mass of water is 18, and the average molecular mass of air is 28.97.

Figure 26.2 *A bottle, sachets and a watch glass containing silica gel, which is a highly effective absorbent material.*

Have you noticed that boxes of vitamin tablets, and even special equipment and leather products, such as bags and purses, often contain sachets of desiccant (drying agent)? Moisture in the air can make food products stale, and leather products can turn mouldy if the humidity is high. The desiccant is usually beads or granules of silica gel, and often a colour indicator, such as cobalt(II) chloride is added to the beads. In Unit 21 you used cobalt chloride paper to test for the presence of water. Silica gel stained with cobalt chloride is blue when dry, and turns pink when it has absorbed moisture from the air. It can be dried out for reuse by heating it in an oven.

Hygroscopic substances

Take a look inside an opened bottle of magnesium chloride, zinc chloride or calcium chloride. It is likely that the granules or crystals are wet. This is because substances like these are **hygroscopic**, which means that they attract water vapour from the air. Desiccants, such as silica gel, are also hygroscopic.

Figure 26.3 *Dry (above) and wet (below) magnesium chloride ($MgCl_2$).*

Assessment

Write a paragraph explaining how you could predict whether today is a humid day or not. How could you test this?

OR

Research the reverse swing of a cricket ball. Is this related to humidity?

27

The water cycle

Keywords
condensation, evaporation, precipitation, stomata, transpiration, water cycle

Did you know?
Clouds can hold enormous amounts of water, with 10 mm of rainfall being the equivalent of 20 000 million tonnes of water.

Where did Earth's water come from? It is most likely that volcanic eruptions at the time of a primitive Earth formed thick clouds of water vapour in the atmosphere. Once Earth's surface cooled, the water condensed and fell as torrential rains that collected in low-lying areas to form the sea.

The sea is the main reservoir of Earth's water supply, but the water is salty. Fresh water comes not only from the municipal services, but from the water cycle. It is through the **water cycle** that the water in our environment is cycled between its liquid, gaseous (water vapour) and even solid (in the form of ice, hail, snow) states. Let us now look at this in more detail in Figure 27.1.

How the water cycle works

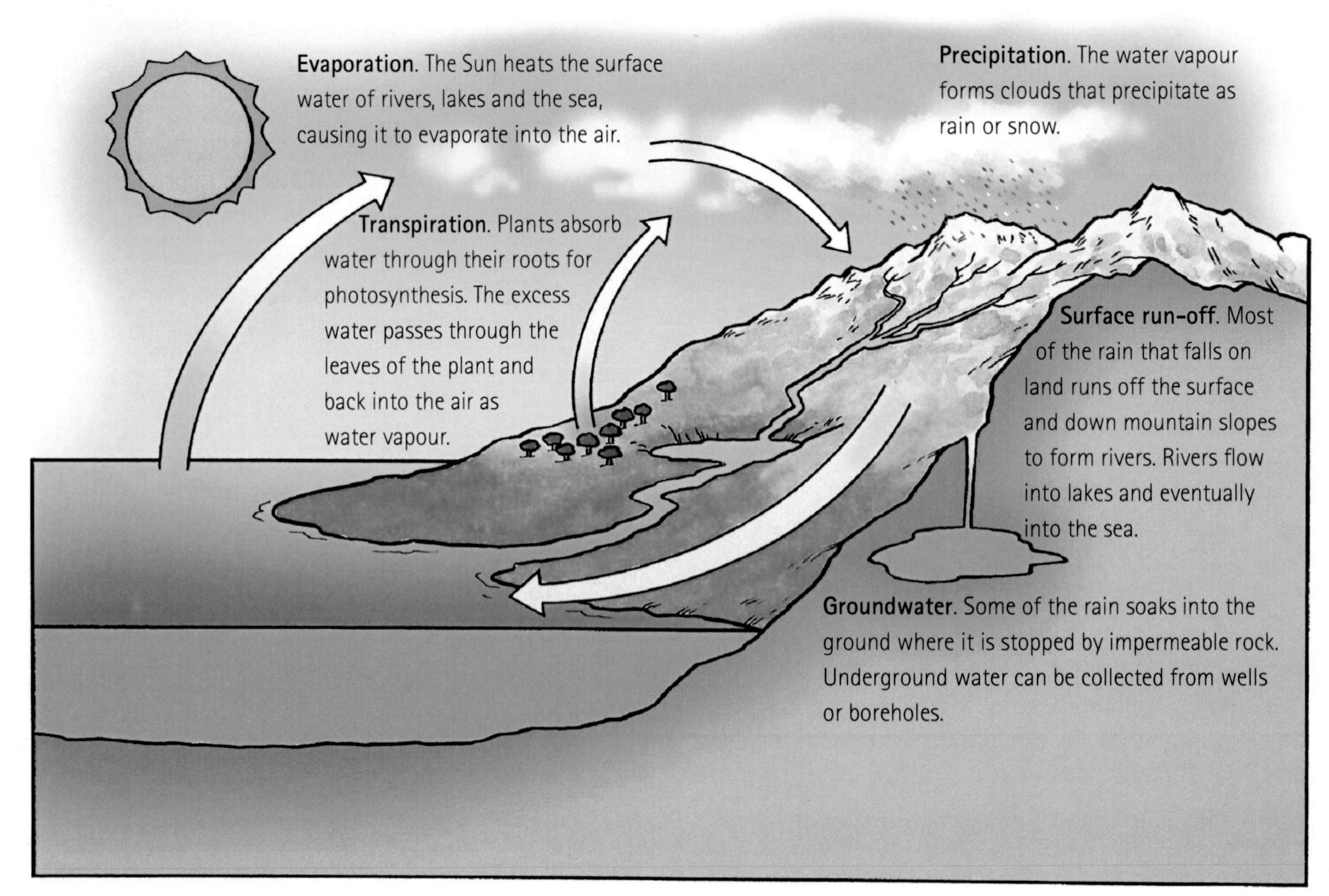

Figure 27.1 *The water cycle.*

ACTIVITY

Although Earth's water supply stays more or less constant, there are local floods and droughts, abundances and shortages. The storage and availability of fresh, unpolluted drinking water often poses a challenge to communities.

1 Find out about the main sources of water in Trinidad and Tobago.
2 Has the country experienced any recent water shortages due to high water demands (e.g. in the tourist season), or incidents of contamination of the water supply?
3 What factors place water resources under threat?

Evaporation and transpiration, condensation and precipitation

You have already, in Year 1, learnt about the water cycle as a global evaporation and condensation process. **Evaporation** and **transpiration** are the processes by which liquid water becomes gaseous water vapour. **Condensation** is the reverse of this process, and **precipitation** is the process by which water returns from the atmosphere to Earth.

Of these, you are probably least familiar with the process of transpiration. More than 90% of the water taken up by the roots of a plant is released into the air as water vapour, most of this through **stomata**, which are the small pores or openings on the surface of the leaves. Gases also enter and leave the plant through the stomata.

Science Extra

Water has an unusual physical property. Solid water (ice) is less dense than liquid water. This means that ice floats on sea water. The layer of ice insulates the water underneath and stops rivers and seas from freezing over completely.

INVESTIGATION Do plants transpire?

1 Put a plastic bag around a well-watered potted plant. Place the plant in a warm, sunny place such as the windowsill, and have a look at the plant the following day. What can you observe?
2 Answer the following:
   a Name the process by which plants lose water through their leaves.
   b What is the name of the pores through which the water is lost?
   c The water vapour from the plant is visible because it has changed to liquid by the process of …
3 Predict: If the same pot plant were placed somewhere cool and shady, would the amount of water collected be the same, less or more?

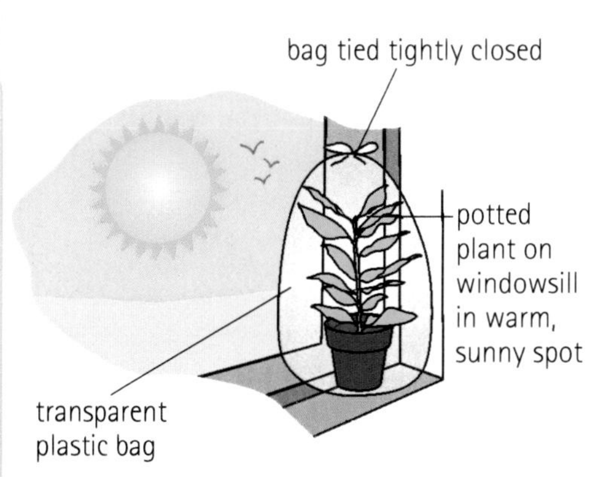

Figure 27.2 *Investigating transpiration in a plant.*

Assessment

In this topic you have learnt about three important cycles – the carbon cycle, the nitrogen cycle and the water cycle. All these cycles are interlinked with one another. Working in groups, have another look at the cycles, and decide how they are interlinked. Each group can present a short play to illustrate how they have interpreted these cycles.

28

Air pollution – how clean is the air?

Keywords

air pollution, air quality, pollutant

Did you know?

Some plant species such as mosses and lichens are very sensitive to pollution. They act as natural indicators or monitors of pollution, growing only in areas where the air is sufficiently clean. The World Health Organisation (WHO) estimates that about 4.6 million people die each year from the effects of air pollution.

What is air pollution?

Air pollution is a common sight in Trinidad. **Air pollution** is the release of any chemical or substance into the air that causes a change to the natural characteristics of the atmosphere. It has the potential to cause harm to plants, animals and the environment.

The biggest source of air pollution is transportation – cars, trucks and aeroplanes – as well as those power plants and refineries that burn fossil fuels. Some of the major air **pollutants** include the following (which you will learn more about in the subsequent units):

Figure 28.1 *Air pollution from a factory chimney.*

Figure 28.2 *Some major air pollutants.*

1. Particulate matter, consisting of small solid or liquid particles, such as those in smoke or spray. The source of these is car exhausts and forest fires.
2. Sulphur dioxide. This gas is produced mostly from the burning of coal and oil in power stations. Sulphur dioxide is the main source of acid rain.
3. Nitrogen dioxide. This gas is produced mainly by car exhausts, and is also responsible for acid rain.
4. Chlorofluorocarbons (CFCs). These gases are used in aerosols, fridges and air conditioning units. They are responsible for damaging the ozone layer that protects Earth from harmful radiation.
5. Volatile organic compounds (VOCs). These are chemicals that are found in many household products such as solvents, paints, cleaning products, aerosols, disinfectants, pesticides, ink, glue, etc. They are also used in industrial processes. They can cause different health problems, ranging from a simple headache to cancer.

ACTIVITIES

1 A few years ago the people in a village near to an asphalt processing plant complained about symptoms such as irritated eyes and headaches. Chemicals leaking from part of the plant were found to be responsible for the symptoms. Research (e.g. on the Internet) the different types of air pollution that this asphalt plant could have produced.
2 Have a look at online Internet sites that display daily **air quality** data, e.g. www.iosoft.co.uk/ccc/ccc.php. Look at the bar graph showing nitrogen oxide levels monitored in Regent Street, Cambridge, UK. What trend do you observe?

INVESTIGATION How clean is the air that you breathe?

Set up the apparatus as shown in the diagram.
1 Tightly plug a piece of cotton wool in the glass tube.
2 Attach the tube to a water suction pump. This will suck air through the cotton wool, which acts as a filter.

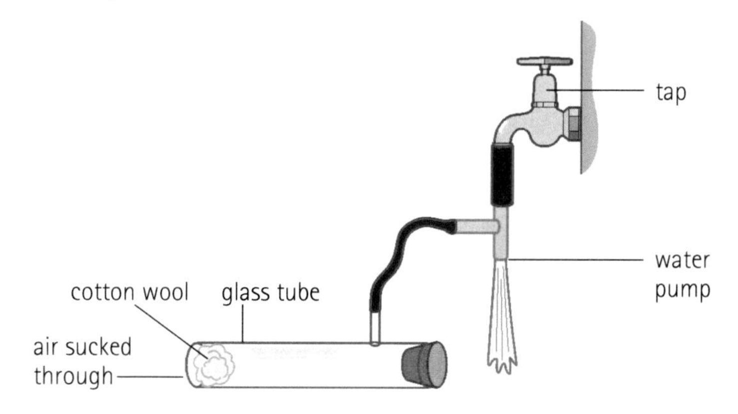

Figure 28.3 *Apparatus for investigating how clean the air is.*

3 Leave the pump running for a few days and then observe your filter.
   - Can you see any change in the colour of the cotton wool? (You can photograph the cotton wool before and after, or compare the used cotton wool filter with a clean piece of cotton wool.)
   - What type of air pollution has your filter trapped and what are the possible sources? (You will learn more about this in the following unit.)
   - Explain how your nose partly filters the air that you breathe.

Assessment

Research why people seek out help from hospitals and clinics more often during the dry season than during the wet season, for ailments affecting the respiratory system.

Science Extra

Lead used to be added to petrol to make car engines run smoothly. Leaded petrol is now banned in most countries after it was found to have harmful effects. Lead poisoning can damage the nervous system, particularly that of children.

29

Particulate pollution and global dimming

Keywords

global dimming, particulate pollution

Did you know?

The average amount of sunlight reaching Earth has fallen by about 6% since 1960.

Particulate matter (PM) – how big or small?

Particles of solid matter, such as soot, ash, metals, dust, and even pollen, are examples of **particulate pollution**. Aerosols (suspensions of tiny liquid particles in the air) are also a form of particulate pollution. Generally, particles in the air that are smaller than 10 μm (i.e. 10×10^{-6} m, which is less than one-tenth the diameter of a human hair) are the greatest threat to human health. These particles are too small to be filtered by the hairs in our nose, or by the mucus lining of our throat and trachea (windpipe), and are carried straight to the lungs.

The effects of particulate pollution

Particulate matter pollution can cause respiratory problems, such as coughing and bronchitis, and can aggravate asthma.

My name is Brian, and I suffer from asthma. Every dry season when there are bush fires and when crops are burnt, my asthma acts up. If I am not careful I have to spend a couple of hours of my busy schedule in hospital being treated for breathing difficulty.

Figure 29.1 *Brian with his asthma pump.*

ACTIVITY

1. List five sources of particulate pollution in the atmosphere.
2. Assess in what ways you or your family contribute to particulate air pollution.
3. What fraction of one millimetre is 10 μm?

In addition to the health hazards, particulate pollution may contribute to potential global problems such as the one described below.

Pan evaporation data – the puzzle

In Unit 27 you studied the water cycle. One of the processes in this cycle is evaporation. As Earth has warmed up over the past 50 years, scientists expected the rate of evaporation from open bodies of water to increase. However, the rate of evaporation in most parts of the world has in fact slowed down. This is probably because less sunlight is reaching Earth than before, due to a phenomenon called **global dimming**.

Science Extra

The greenhouse effect has been making the world hotter (see Unit 32). On the other hand, global dimming reduces how much sunlight reaches the Earth. If we reduce particulate pollution, global dimming should decrease, but this might lead to an even bigger increase in global warming.

How particulate pollution causes global dimming

Particulate pollution could be causing global dimming in two ways:

It reduces the amount of sunlight that can reach the Earth.

The particles in the air act like a filter, reducing the amount of light that can pass through.

It increases the amount of sunlight being reflected back from the clouds into space.

Raindrops form when water vapour condenses around small particles in the air. A large amount of particulate pollution results in the formation of more small raindrops, rather than fewer large ones. Raindrops act like a mirror and the small raindrops reflect more light than the larger ones.

Assessment

Dirt on walls, windows and roofs of buildings is one of the obvious signs of particulate air pollution. Make a note of dirty buildings at school or near where you live. For example, are the walls that face the street dirtier than other parts of the building?

30

Nitrogen oxides, sulphur oxides and acid rain

Keywords

acid rain, nitrogen dioxide, nitrogen monoxide, sulphur dioxide

Many industries, particularly power stations, burn a lot of coal and oil to make electricity. These fossil fuels contain sulphur and nitrogen impurities, and so their combustion produces gaseous sulphur dioxide (SO_2) and nitrogen dioxide (NO_2).

Nitrogen oxide – the brown colour of smog

First, **nitrogen monoxide** (NO) is formed as a by-product of combustion. It is formed when fuel containing nitrogen is burned, or when nitrogen in the air reacts with oxygen at very high temperatures:

$$N_2 + O_2 \rightarrow 2NO$$

It is not only emitted from fossil-fuel-burning power plants, but also by car exhausts. Nitrogen monoxide then reacts with oxygen in the air to produce nitrogen dioxide:

$$2NO + O_2 \rightarrow 2NO_2$$

Nitrogen dioxide is a brown gas that gives smog its dirty colour.

Sulphur dioxide – the smell of burnt matches

By contrast, **sulphur dioxide** is a colourless gas. At high concentration it smells like burnt matches or rotten eggs. Both sulphur dioxide and nitrogen dioxide gases are lung and eye irritants.

Science Extra

pH is a measure of how acid (or alkaline) a solution is. The pH scale goes from 1 to 14. A neutral solution has a pH of 7. Anything above 7 is alkaline, and anything below 7 is acidic. Rain water is naturally slightly acidic; it has a pH of about 6. Vinegar has a pH of about 4.

How acid rain forms

Both sulphur dioxide and nitrogen dioxide gases combine with atmospheric water and oxygen in the following reactions to produce sulphuric acid and nitric acid, which then fall as **acid rain**:

$$2SO_2 + O_2 + 2H_2O \rightarrow 2H_2SO_4$$
$$4NO_2 + O_2 + 2H_2O \rightarrow 4HNO_3$$

Sulphur dioxide is responsible for more than 70% of acid rain.

Figure 30.1 *The formation of acid rain.*

Simulating the formation of acid rain

Your teacher will demonstrate this experiment for you.

1. Place a piece of sulphur on a combustion spoon.
2. Carefully light the sulphur and immediately place it into a gas jar with some water in it.
3. Observe what happens as the sulphur burns.
4. After it has finished burning, shake the gas jar so that the gas dissolves in the water.
5. Carefully open the lid and put in a few drops of universal indicator. What colour does the universal indicator turn? What is the acidity of the water?
6. Compare this with the same volume of tap water to which the universal indicator has been added.

The effects of acid rain on the environment

Acid rain accelerates the weathering of carbonate rocks, resulting in the acidification of rivers and streams, which in turn damages forests. The uptake of acidic water by trees reduces their ability to grow and their resistance to disease. Also, acidic water in the soil can leach it of important nutrients.

Figure 30.2 *These trees are dying because of acid rain.*

Ways of reducing SO_2 emissions and acid rain

Many coal-burning power plants now have devices called 'scrubbers' on their chimneys that reduce the amount of sulphur dioxide they release into the air. The scrubber filter is injected with a lime slurry (calcium carbonate) that reacts with the sulphur dioxide to form calcium sulphate (gypsum), which has a neutral pH. Also, 50% of the sulphur can be removed from coal by crushing and washing it before burning it.

The Clean Air Act in the United States aims to reduce sulphur oxide emissions by 2010 to half their 1980 levels. Power plants have been allocated allowances or quotas for the amount of SO_2 they can release into the air each year. These allowances can be bought and sold among power plants. A power plant that makes use of low-sulphur coal, and that has reduced its emissions below the limit, can sell its unused quota to compensate for the extra cost of the low-sulphur coal.

Figure 30.3 *United States Environmental Protection Agency (EPA) projections of SO_2 emissions.*

Ways of reversing the damage

Acidic water in lakes and streams can be neutralised by adding alkaline substances such as limestone.

Assessment

Write a short paragraph to explain what the graph in Figure 30.3 tells us.

Ozone – protectant and pollutant

Keywords

allotrope, catalyst, CFCs, ozone, stratosphere, ultraviolet light

Did you know?

Unlike ozone molecules, chlorofluorocarbon (CFC) molecules are very stable. It is estimated that a CFC molecule released into the air takes 15 years to reach the stratosphere. Once there, it can last for a hundred years, and can destroy as many as 100 000 ozone molecules.

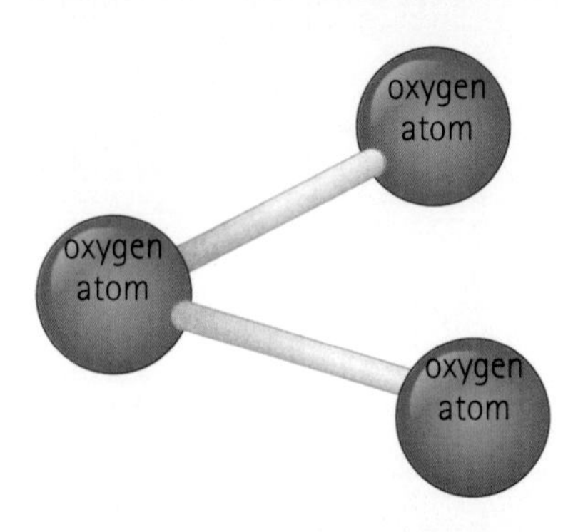

Figure 31.2 *A model of an ozone molecule.*

Ozone (O_3) is a gas whose molecules are made up of three oxygen atoms, and is an **allotrope** (different physical form) of the element oxygen. It is mainly produced high up in the atmosphere when **ultraviolet light** (UV light) reacts with oxygen. UV light splits oxygen molecules into oxygen atoms. These oxygen atoms then combine with other oxygen molecules to form ozone.

$$O_2 \xrightarrow{UV} 2O \quad \text{and} \quad O + O_2 \rightarrow O_3$$

The stratospheric ozone layer – our ozone security system

A thin layer of ozone in the upper atmosphere or **stratosphere** (about 50 km above Earth's surface) surrounds Earth. This layer plays an important role in protecting Earth by filtering out most of the Sun's harmful ultraviolet radiation. Too much exposure to UV radiation can damage DNA and cause skin cancer, especially in light-skinned people who have less natural protection.

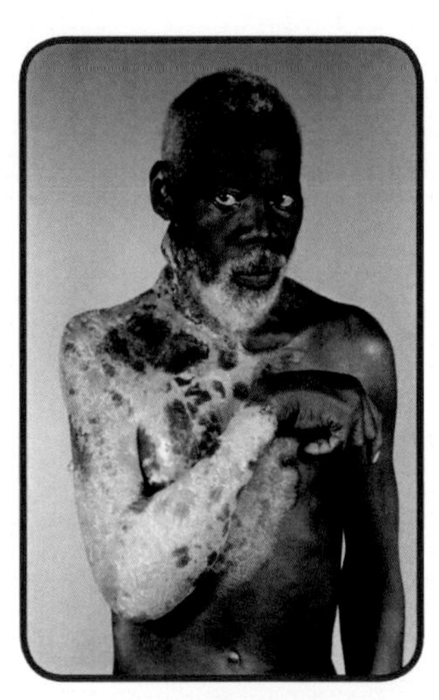

Figure 31.1 *A person suffering from skin cancer caused by UV radiation.*

The hole in the ozone layer

In 1985 scientists noticed that a hole was developing in the ozone layer over the polar regions, which meant that the amount of ozone in the atmosphere was becoming depleted. Ozone depletion also refers to a decrease in the total amount of ozone in the stratosphere. With depletion of the ozone layer comes the risk of increased exposure to UV rays.

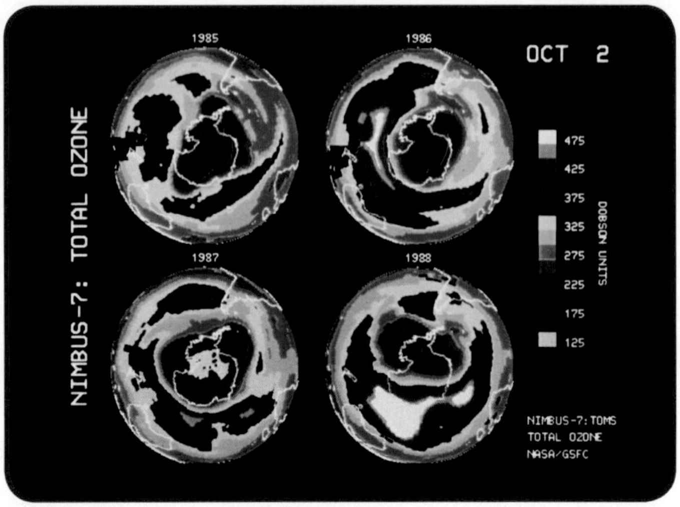

Figure 31.3 *These satellite maps show the variation in the ozone over Antarctica from 1985 to 1988. The lowest ozone levels (i.e. the ozone hole) are shown in pink and purple, with 1987 (bottom left map) showing a record low. The size of the ozone hole varies from year to year. It is caused by pollution in the atmosphere caused by CFCs from aerosols and refrigerators.*

Ozone-depleting substances (ODSs)

Chlorofluorocarbons (CFCs) are mainly responsible for ozone depletion. **CFCs** are a group of manufactured, volatile chemicals or chlorine-containing gases, invented in the 1920s. They were used as propellants in aerosol sprays such as deodorants, and are still used today as coolants in fridges and air conditioning systems. CFCs are long-living molecules. Released into the air, they enter the stratosphere, and are broken down into chlorine atoms, which act as catalysts. A **catalyst** is an agent that speeds up a chemical reaction, but itself remains unchanged by the process.

A chlorine atom reacts with an ozone molecule, taking an oxygen atom to form chlorine oxide and molecular oxygen:

$$Cl + O_3 \rightarrow ClO + O_2$$

The free oxygen atom then takes the oxygen from the chlorine oxide, forming molecular oxygen, and freeing the chlorine atom again, to be recycled.

If you cancel out, the overall reaction, in which ozone is destroyed by the reaction with atomic oxygen, is:

$$O_3 + O \rightarrow 2O_2$$

CFCs are no longer used in most aerosols, and the hole in the ozone layer is reducing slowly. The rate at which the ozone layer was thinning seems to have slowed down. But we still use harmful CFCs. It is estimated that Trinidad and Tobago releases almost 100 tonnes of CFCs into the atmosphere every year. Most of this (82%) comes from discarded fridges and air conditioning units. Everyone has a duty not to pollute the atmosphere that we all share. Freon, which contains CFCs, should be removed from discarded refrigerator and air conditioning units by a qualified person.

Science Extra

Bromofluorocarbon compounds are known as halons. One bromine atom can destroy about 10 times as much ozone as a chlorine atom.

Did you know that the Trinidad and Tobago Solid Waste Management Company Limited (SWMCOL) will take away and dispose of large items such as fridges?

ACTIVITY

Discarded fridges release damaging CFCs into the atmosphere when they rust. Create a poster telling people why they should not illegally dump their old fridges.

Ozone and smog in the trophosphere

Although ozone is important in the stratosphere, ground-level ozone is a pollutant. It is formed by the reaction of sunlight on air that contains hydrocarbons and nitrogen oxide pollutants. This is nature's way of cleaning up the air, since ozone is a natural purification agent. It is usually broken down quickly, but it can build up near and downwind of urban and industrial areas. Ozone mixes with other pollutants in the air to form smog, which reduces visibility and can irritate the eyes and lungs. Hot weather encourages the formation of smog.

Assessment

Explain how a person using an aerosol in Trinidad might contribute to someone in Norway getting skin cancer.

32 The greenhouse effect and global warming

Keywords
global warming, greenhouse effect, greenhouse gases

The greenhouse effect

Gases such as carbon dioxide, water vapour and ozone in the atmosphere help to keep the surface of Earth warm. They are sometimes referred to as **greenhouse gases** because, like the glass of a greenhouse, they trap in the heat (although a greenhouse does this by preventing convection, whereas the greenhouse gases insulate).

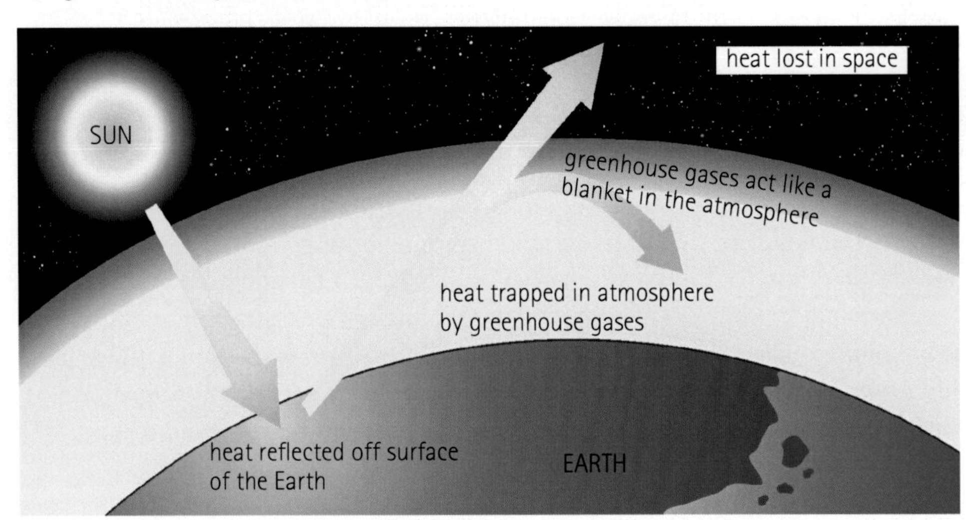

Figure 32.2 *How the greenhouse effect works.*

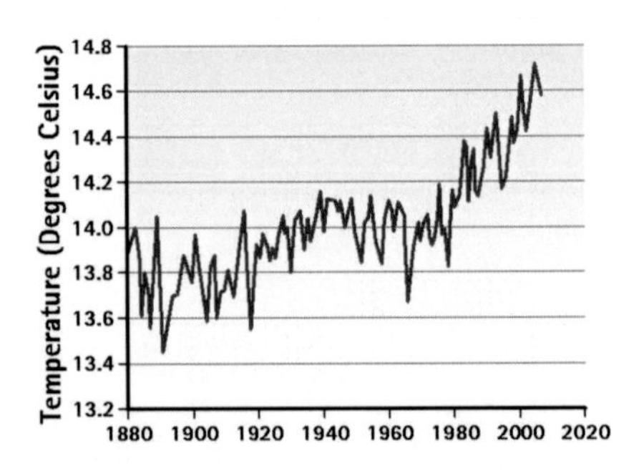

Figure 32.1 *Global temperature increases.*

Some of the radiation from the Sun hitting Earth is absorbed, and some of it is reflected from the surface of Earth back into space. Greenhouse gases in the atmosphere prevent some of the reflected heat from escaping. This is called the **greenhouse effect**. If it weren't for this insulation, the average surface temperature of Earth would be 30 °C lower than it is.

An increase in greenhouse gas emissions contributes to global warming

With increased human activity and industrialisation – the burning of fossil fuels and deforestation – the amount of greenhouse gases, particularly carbon dioxide, released into the atmosphere has been increasing. More heat is trapped by this thicker carbon dioxide blanket, and so Earth is becoming hotter. Global temperatures have increased by about 0.5 °C in the last century (see Figure 32.1). This phenomenon is called **global warming**. An average temperature increase of 1 °C has been measured for the Caribbean between 1976 and 2000. The temperature increases projected for the future in the Caribbean are shown in Table 32.1.

	Temperature increase (°C) Scenario 1 (low)	Temperature increase (°C) Scenario 2 (high)
Dec–Feb		
2050	1.4	2.0
2080	2.0	3.3
June–Aug		
2050	1.5	1.9
2080	2.0	3.3

Table 32.1 *Projected temperature increases for the Caribbean for 2050 and 2080 – winter and summer.*

The world's climate is changing

Although these temperature increases seem small, an average increase of temperature between 1.5 and 6 °C is enough to cause a significant change in climate. With changes in temperature come changes in rainfall patterns, the melting of the Arctic ice cap (accompanied by rising sea levels), and increased hurricane activity.

Science Extra

Most countries of the world signed the 'Kyoto Protocol' in 2006. This is an agreement to reduce greenhouse gas production. However, countries such as Australia and the US have not yet agreed to the treaty.

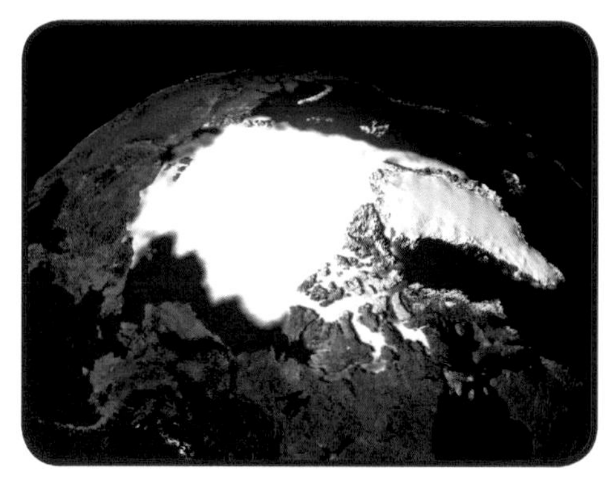

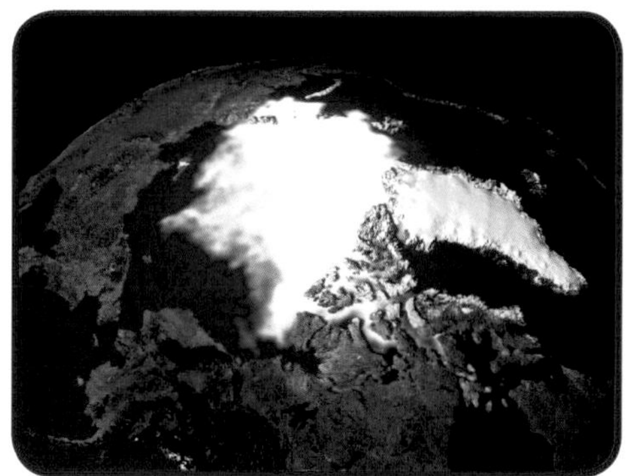

Figure 32.3 *These satellite pictures show the Arctic Sea ice in 1979 (left) and 2003 (right). A comparison between the pictures shows how global warming since 1979 is melting the sea ice around the North Pole.*

The number of hurricanes and tropical storms in the Atlantic-Caribbean basin has increased from seven to ten a year since 1886, and the sea level has risen by up to 20 cm in the last century.

Figure 32.4 *Increased hurricane activity is being blamed on climate change.*

The impact of climate change

As a cluster of islands, the Caribbean is particularly vulnerable to rises in sea level and sea surges following tropical storms. Salt water intrusion has already been detected in the west coast aquifers in Barbados, for example, with the water supply in these underground aquifers gradually becoming salty. A rise in sea level will lead to salt water contamination of the soil and fresh water resources. Tropical storms and increased hurricane intensity can result in damage to the habitat, increased flooding and landslides, outbreaks of disease, and loss of biodiversity.

ACTIVITY

Explain the meaning of these terms:
greenhouse effect, global warming, climate change

Assessment

A good source of climate change data is elderly people, who have experienced the climate changes at first hand. Do a survey of some older people and ask them if they have noticed climate changes over the years.

33

Air pollution – responsibility and change

Keywords

catalytic converter

Did you know?

Catalytic converters are connected to the end of vehicle exhaust systems. Although a catalytic converter prevents a large amount of the harmful gases from escaping into the air, a sizeable amount of these gases is still released into the environment.

Who is responsible for reducing air pollution?

By now, you should have an understanding of the importance of Earth's air and atmosphere, and of the problems (environmental and health-related) caused by air pollution. But who is responsible for dealing with this complex issue? What can we as individuals do? What sort of lifestyle changes or sacrifices must be made? Environmental problems are everyone's responsibility. Different people have different responsibilities, but everyone can do something.

ACTIVITY

1 Look at these two examples:

Many people drive cars that pollute the air unnecessarily. There are regulations about emissions from cars, and people should get a licence to show that their cars are within legal emissions limits. But some drivers might get a licence from a friend, without getting their car fixed or modified.	Concrete is not a good insulator, so our houses become hot. We then use air conditioning systems to cool them down, but this uses a lot of energy and pollutes the air, contributing to the depletion of the ozone layer.

As a group or a class, discuss what responsibilities each of the following groups have to solve the two example problems:

- ordinary people
- industry
- government (national and international).

2 We all know that cars pollute the air, but who would choose not to have a car because of pollution? Is it human nature to think one thing and then do another? Why don't people do the environmentally friendly thing? When answering these questions, consider factors such as financial constraints, education and attitude. What do you think would make people change their behaviour?

3 Write up your discussions for a newspaper article, or summarise your discussions on the chalkboard or a large sheet of paper.

Cars and cleaner fuels

Figure 33.1 *Running a car on CNG autogas can be good for your wallet and your environment.*

Science Extra

The cleanest fuel of all is hydrogen, since the only product of its combustion is water. Some car manufacturers are trying to develop cars that run on hydrogen so that we can completely eliminate pollution from car emissions.

As people become more aware of the pollution caused by car exhausts they look for cleaner fuel alternatives. In Trinidad, where we produce compressed natural gas, CNG is an alternative. Any petrol car can be converted to run on CNG by installing a CNG tank. Although the CNG gas is much cheaper than petrol or diesel, there is the initial cost of the conversion to consider.

In Brazil, engineers are exploring the feasibility of running cars on ethanol made from sugar cane. Pure alcohol is a very clean fuel.

Catalytic converters

Car exhausts are one of biggest causes of air pollution. To help reduce the toxicity of emissions, most cars are now fitted with **catalytic converters** in their exhaust systems. Three-way catalytic converters convert:

- nitrogen oxide to nitrogen and oxygen,
- carbon monoxide to carbon dioxide, and
- unburnt hydrocarbons to carbon dioxide and water.

catalytic converter

Figure 33.2 *Catalytic converters like this one reduce pollution from car exhausts.*

Assessment

Interview people about their cars and their awareness about how their cars contribute to air pollution. Include someone who has converted their car to CNG use. Ask about the advantages and disadvantages of this fuel. Research the latest trends in car technology (e.g. electric cars) aimed at reducing emissions and the reliance on fossil fuels.

34

Solutions to air pollution

Keywords
technology

Did you know?

Scientists at the University of the West Indies (UWI) have been developing a solar fridge. This is a fridge that will use sunlight to evaporate the coolant that keeps a fridge cold.

Nature's way of cleaning the air

You have already seen how ozone can work as a natural air purification agent, provided the air pollution isn't too concentrated. Winds and water (in the form of rain) are also nature's brooms and brushes for cleaning up the air. Although Trinidad is a highly industrialised island, we do not suffer too badly from air pollution. Much of our industry is located on the south and west coasts of Trinidad, and our prevailing winds come from the north-east, carrying the pollution out to sea. It is important to bear in mind, though, that often the problem is simply transported somewhere else.

Figure 34.1 *A view over San Fernando on a bad air-pollution day.*

Falling rain also helps to clean the air. If, however, the pollutants are sulphur dioxide and nitrogen dioxide, and their concentrations are high, the rain will fall as acid rain. We cannot therefore rely on natural processes to fix our air-pollution problems.

Science Extra

Before decisions are made by governments about a proposed development, they carry out an Environmental Impact Assessment (EIA), which aims to predict what effect the development would have on the local environment.

Innovative ways of solving the air pollution problem

Together with aiming to reduce air pollution (see Unit 33), we can use **technology** to address the air-pollution problem. Read the article on page 73 as an example.

ACTIVITY

Read the article on the next page and then answer these questions.

1 What is the name of this anti-air-pollution technology, and what is the active ingredient?
2 Describe how the product works.
3 Explain what a catalyst is.
4 Which air pollutants does it target?
5 What are the advantages and disadvantages of the product?

'A concrete step toward cleaner air'

Visitors to the Italian Pavilion of the architecture exhibition in the Venice Biennale will get a breath of fresh air. That's because parts of the concrete walls and grounds have been built with cement containing an active agent that, in the presence of light, breaks down air pollutants such as carbon monoxide, nitrogen oxide, benzene and others through a natural chemical process called photocatalysis.

The demonstration is a reminder that smart innovation can offer unexpected solutions even for complex problems such as air pollution. The technology, called TX Active, has been under development for almost 10 years in the labs of Italcementi, the world's fifth-biggest cement producer, and is starting to be applied commercially to buildings and streets in Italy, France, Belgium and elsewhere.

The results so far are astonishing: A street in the town of Segrate, near Milan, with an average traffic figure of 1 000 cars per hour, has been repaved with the compound, '... and we have measured a reduction in nitric oxides of around 60%', says Italcementi's spokesperson Alberto Ghisalberti. In large cities such as Milan (with persistent pollution problems caused by car emissions, smoke from heating systems, and industrial activities), both the company and outside experts estimate that covering 15% of all visible urban surfaces (painting the walls, repaving the roads) with products containing TX Active could abate pollution by up to 50%, depending on the specific atmospheric conditions.

Of course, this approach isn't meant to replace efforts to curb pollution, but it can significantly magnify their effects. Here's how it works: The active ingredient – basically a blend of titanium dioxide that acts as a photocatalyser – can be incorporated in cement, mortar, paints and plaster.

In the presence of natural or artificial light (this applies also indoors) the photocatalyser significantly speeds up the natural oxidation processes that cause the decomposition of pollutants, transforming them into less harmful compounds such as water, nitrates or carbon dioxide.

'These aren't necessarily "clean", but from an environmental standpoint they're much more tolerable,' says Rossano Amadelli of the Italian National Research Council (CNR), the scientists who led the laboratory testing of the TX Active materials.

The patented pollution-reduction technology comes at a premium, of course, but the extra cost is limited by the fact that the active ingredient only needs to be used on the surface.

Photocatalytic blocks cost about one-third more than normal paving blocks, but this is still far cheaper than the long-term cost of doing nothing about air pollution.

It turns out that the photocatalysing cement has another advantage, one that has great appeal to star architects such as Richard Meier. TX Active not only hastens the decomposition of organic and inorganic pollutants, it also prevents their build-up on surfaces, helping to preserve a building's pristine appearance over time.

Source: Bruno Giussani, Business Week, *5 November 2006 (adapted)*

Assessment

What would be your solution to the air-pollution problem? Present your idea as a one-page article, or oral presentation.

Review topic 2

The air in our atmosphere is a mixture of gases that support life. The atmosphere is a complex and dynamic system, in which the component gases such as oxygen, nitrogen, carbon dioxide and water vapour are recycled. It can also be thought of as a 'chemical factory' in which physical and chemical changes take place. Pollution of the air disrupts the balance of the system, and introduces substances that can harm human health and the living environment. Some air pollutants are of natural origin, but the air pollution problem is caused by humans. There are some innovative ways of addressing the problem and it is in our interest (short-term and long-term) to reduce air pollution.

Multiple-choice questions

1 What is the most abundant gas in the atmosphere?
   a oxygen
   b water vapour
   c nitrogen
   d carbon dioxide
2 Which one of the following gases is used in manufacturing coloured advertising lights?
   a oxygen
   b nitrogen
   c neon
   d carbon dioxide
3 Which of the following gases produces acid rain?
   a oxygen
   b neon
   c ozone
   d sulphur dioxide

Longer questions

1 Study the table below before answering the questions that follow.

Components of air	Exhaled air (%)	Inhaled air (%)
Nitrogen	79%	79%
Oxygen	21%	16%
Carbon dioxide	0.03%	4%
Other gases	1%	1%

   a Draw a pie chart to show the data for exhaled air.
   b Account for the differences between exhaled and inhaled air.
   c Describe the test for oxygen.
   d Describe the test for carbon dioxide.
   e Which of the gases in the table is responsible for global warming?
2 Write a short paragraph to summarise the carbon cycle, explaining how plants and animals are dependent on one another.
3 a Sketch the nitrogen cycle.
   b Explain the importance of lightning in the cycle.
   c Explain the importance of nitrogen-fixing bacteria in the cycle.
4 By which of these processes is water put into the atmosphere:
evaporation, precipitation, transpiration?
5 'Fossil fuels are responsible for polluting our environment and the increase in global warming.'
   a Name the gases produced that are responsible for the effects referred to in the above statement.
   b What is acid rain? Give three examples of acid rain damage in Trinidad and Tobago.
   c What can we do to minimise or solve our acid rain problem?
   d Should small island states like Trinidad and Tobago be part of the solution to stop global warming?
6 Use the information in the table below to answer the questions that follow.

Year	Concentration of carbon dioxide (p.p.m.)	Average temperature (°C)
1959	305	26
1993	340	28
2005	378	31

   a Plot a bar graph showing the concentration of carbon dioxide and average temperature for each year.
   b Discuss the relationship between carbon dioxide levels and average temperature.
   c Suggest three actions that can be taken to reduce carbon dioxide emissions.
7 An international company wants to set up an industrial plant in Trinidad and Tobago. One of the wastes generated by this plant is an airborne dust.
   a What are the possible effects of this waste on the environment and humans?
   b Give suggestions to the builders of this plant on how to minimise the effects of this waste and so produce a viable industrial complex.
8 Design a cartoon showing the effects of global warming.
9 a State five ways of reducing the effects of global warming.
   b Discuss two of these ways in detail.
10 What are the effects of the annual occurrence of Saharan dust on the Caribbean?

Our place in space

Keywords

astronomy, galaxy, light year, Milky Way, universe

Did you know?

Galaxy means *milky circle*. Three types of galaxy have been identified, according to their shape – elliptical, spiral or irregular.

Astronomy is probably one of the oldest natural sciences, but it was only in the 16th century that Nicolaus Copernicus put forward the revolutionary theory that Earth rotated on its axis once a day and that it travelled around the Sun once a year. Until the 1500s it was generally accepted that Earth was at the centre of the solar system and that all the planets revolved around Earth. Thanks to the work of early astronomers (such as Copernicus, Galileo, Newton, Kepler and Herschel) and modern-day scientists, our knowledge and understanding about our solar system and the universe has evolved.

We now know that our planet Earth orbits a star we call the Sun. We call the Sun and orbiting planets the solar system, and we share our solar system with eight other planets (no longer nine, since Pluto has lost its planet status). We know that a **galaxy** is a system of stars held together by gravitational forces and that the Sun is one star in a spiral-shaped galaxy of more than 200 billion stars called the **Milky Way**.

Figure 36.1 *The Milky Way, showing our galaxy's spiral structure.*

Because galaxies are grouped or clustered together, we can add more information to our cosmic address. Our galaxy, the Milky Way, is a member of a group of about 30 galaxies called the 'Local Group'. Of these, the Milky Way and the Andromeda Galaxy, our neighbouring galaxy, are the brightest. The Local Group in turn is part of the Virgo Supercluster of galaxies.

Figure 36.2 *Nicolaus Copernicus was a Polish astronomer who lived from 1473 to 1543.*

The universe – infinite and expanding

How many galaxies are there in the universe? The answer is unknown. There are probably more than 100 billion galaxies in the visible part of the **universe**, but the universe itself is probably infinite in size. This means it has no end or boundaries, and according to the 'Big Bang' theory, the universe is expanding. The 'Big Bang' theory states that the universe was born out of an enormous explosion that took place about 13.7 billion years ago.

The universe or cosmos is a big place. On Earth we measure distance in metres or kilometres, but space is so enormous that it is impractical to use the same units. So instead we measure distance in light years. Light travels very fast – at a speed of almost 3×10^8 m/s. A **light year** is defined as the distance that light would travel in a year, and is equal to approximately 9 500 000 000 000 km. For comparison, the diameter of Earth is about 12 760 km; our Milky Way Galaxy is 100 000 light years across.

ACTIVITY

An expanding universe does not mean that the stars and galaxies are getting bigger, but that the space or distance between them is increasing with time. Also, the further away a galaxy is from us, the faster it moves away. To demonstrate this, work with a partner to make a model of the expanding universe.

1 Mark an uninflated balloon with a cross to represent our Milky Way and mark two more crosses to represent two other galaxies – one near (A) and one far (B).
2 Blow one breath into the balloon, twist and hold it sealed while your partner measures the distance between the 'Milky Way' and each of the two 'galaxies'. (It is difficult to use a ruler on a curved surface, so use a strip of paper and mark up the distance. Then measure this distance against a ruler.)
3 Now blow the balloon up large while your partner counts or times how long it takes you to do this, e.g. 5 seconds. Again, hold the balloon sealed while your partner measures the distances.
4 Now calculate the 'speed' with which the two 'galaxies' moved away from the 'Milky Way'. An example for one galaxy (e.g. galaxy A) is given here.

Distance 1	Distance 2	Speed = (Distance 2 - Distance 1) ÷ time
2 cm	3 cm	(3 - 2) cm ÷ 5 s = 0.2 cm/s

5 Answer the following.
   a Which 'galaxy' moved away faster from the 'Milky Way' – A or B?
   b As the balloon expands, the sizes of the marks you made on the balloon get bigger. As the universe expands, do the galaxies increase in size in this way?

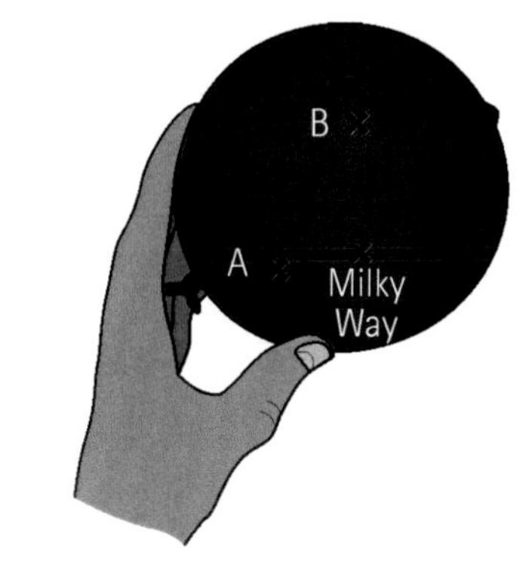

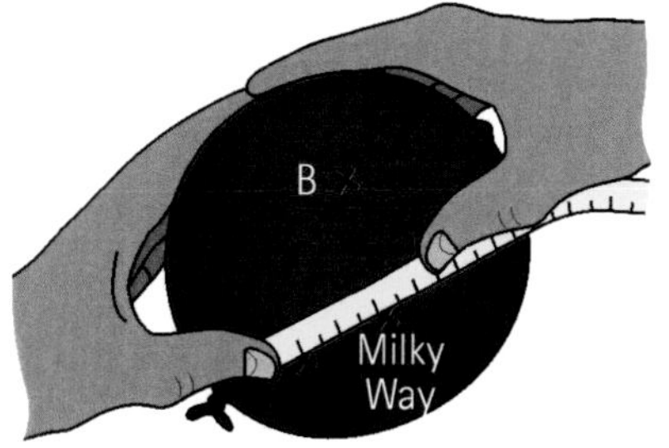

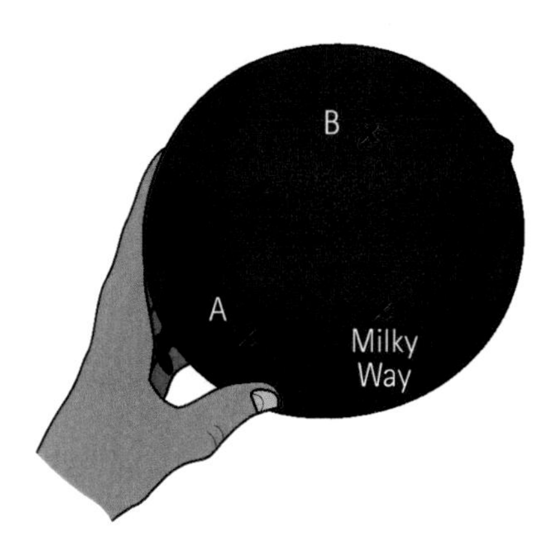

Figure 36.3 *Illustrating the expanding universe using a balloon.*

Assessment

This table lists some of the common words used to study space and gives their meanings. Match each word to its meaning and then copy the table into your book.

Word	Meaning
universe	a natural satellite that orbits around a planet
solar system	a large celestial body that orbits the Sun
star	a massive body in space that produces heat and light
moon	our Sun and the planets that orbit around it
planet	is made up of billions of stars
galaxy	the total matter that exists in all of space

Science Extra

The universe is expanding but scientists predict that eventually it will shrink and collapse. This is referred to as the Big Crunch.

37

Looking into the night sky

Keywords

asterism, constellation, North Star, planet, star

Did you know?

The best way to explore space while staying on Earth is to go to a planetarium. A planetarium is a specially shaped theatre that can project pictures of space onto the ceiling. There is one such planetarium at NIHERST near D'Abadie.

Science Extra

Light is the fastest thing in the universe. Nothing can travel faster than light. When you look at a star in the sky, you are looking at the light that left that star many years ago and has travelled across the universe to us. Some of the stars that you see at night might not even exist anymore.

Stars and planets

When you look up at the night sky, most of the objects you see belong to the Milky Way Galaxy. Of the planets, Venus (sometimes called the 'Morning star' or the 'Evening star') is the easiest to see. When visible, it is the brightest object in the sky, apart from the Sun and the Moon. Mars, Jupiter and Saturn are also easy to see at the right times.

How can you tell the difference between a star and a planet?

Stars are luminous, which means they give off their own light, while **planets** do not. Stars twinkle because their light is bent and reflected by gas and dust in Earth's atmosphere. We see the planets only because of light reflecting off them, and so they are trickier to see.

The nearest and largest planets in our solar system can be seen with the naked eye, and because they orbit the Sun in the same plane, they are always seen positioned in a narrow zone in the sky. These planets are brighter than any of the stars seen in this zone. The first planets outside our solar system were discovered in the 1990s and presently, more than 2 000 extra-solar planets have been identified. HD20945b and OGLE-2005-BLG-390Lb are two of these.

Constellations

Constellations are patterns of stars identified by astronomers long ago, whereas **asterisms** are modern groupings of stars. To find your way around the night sky you need to learn a few landmark star patterns. The table alongside gives you four to begin with, showing you at which time of the year they are visible in the night sky. Orion is a constellation and the other three are asterisms.

Season	Constellation or asterism
winter	Orion
spring	Big Dipper
summer	Summer Triangle
autumn	Great Square of Pegasus

Table 37.1 *Constellations or asterisms you can see during the four seasons*

Figure 37.1 *Orion.*

Figure 37.2 *The Summer Triangle: Cygnus, Lyra and Aquila.*

Why do you think the same star patterns can't be seen from the same place on Earth in the night sky all year round?

The North Star – Polaris

As Earth turns on its axis, the stars move across the night sky.

The image below was photographed during the course of a night and shows how the stars appear to move in the sky. Can you see the one star in the centre of the photograph that does not move? This is the **North Star** or Polaris, the 'pole star', which was used for navigation by sailors long ago. The North Star lies along the same axis that Earth rotates on, so as Earth rotates, the position of the North Star does not change.

Figure 37.3 *Star trails in the night sky with the North Star in the centre.*

Figure 37.4 *The North Star 'stays' in the same place in the night sky as Earth rotates.*

The brightest star and the closest star

The brightest star in our night sky is Sirius, which is part of the Canis Major or Big Dog constellation, and is sometimes called the Dog Star. (See the picture of Sirius on page 83, Figure 39.3.) Our closest star is Proxima Centauri, which is 4.3 light years away. It is one of three stars in the Alpha Centauri cluster. To the naked eye, the cluster is visible as a single star and brightest one in the Centaurus constellation.

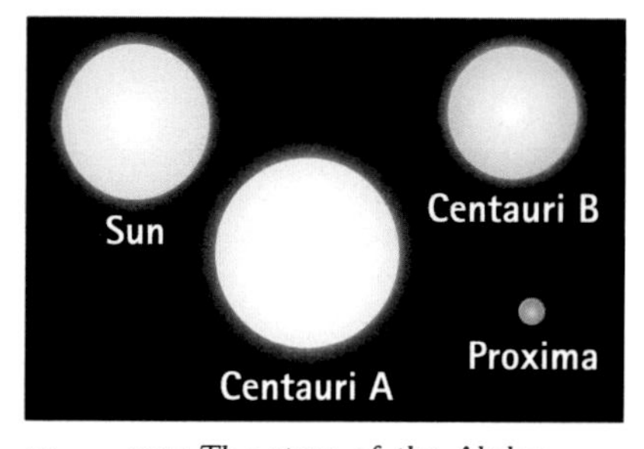

Figure 37.5 *The stars of the Alpha Centauri cluster and their sizes relative to the Sun.*

ACTIVITY

On a clear night, as far away from the city lights as possible, try to identify some of the individual stars, planets and constellations or asterisms you have learnt about. Use a sky map or chart that is accurate for the month, or visit sites such as www.nightskyinfo.com or www.SkyTonight.com to find out which stars and planets should be easy to see. Start by trying to find the North Star. Binoculars or a telescope are not essential for stargazing, but make use of them if you can.

Assessment

Learn the names of three constellations and how to find them in the night sky.

38

The Hubble telescope – a window in space and time

Keywords

black hole, Hubble telescope, NASA

Did you know?

The University of the West Indies' St Augustine campus has a large observatory close to Mount Benedict that is used to study space. Observatories house large telescopes and are built in places away from city lights where the air is clean and the sky is usually clear. The largest telescopes in the world are on the summit of Mauna Kea mountain in Hawaii.

Figure 38.1 *The Space Shuttle Discovery.*

Using telescopes

Telescopes were first developed in the 1600s and allow us to see some objects in space which we otherwise wouldn't be able to see. Titan, the largest moon of Saturn, was one of the first discoveries made with the aid of a telescope in 1655.

Telescopes also help us to see objects in more detail. The bigger the telescope is, the more light is let in and the more detail you can see. You will learn more about making a telescope in the next topic when you study light.

The Hubble telescope – the first telescope in space

Even the observatories' largest telescopes have their limitations – they are still looking at space through Earth's distorting atmosphere. To get better images we need to go beyond the atmosphere. So, in 1990, after 20 years of designing and building, the National Aeronautics and Space Administration (**NASA**) launched the **Hubble telescope** into space using the Space Shuttle Discovery.

Since then, the Hubble telescope, which orbits 600 km above Earth, has relayed information about the birth of stars and the evolution of galaxies. Some of its most significant discoveries include its dating of the universe as 12–14 billion years old (by measuring the distances to 18 galaxies) and confirming that black holes do exist. **Black holes** are collapsed stars whose

Figure 38.2a *An image of the planet Saturn, taken by the Hubble telescope. Two of Saturn's nine moons are visible as white dots.*

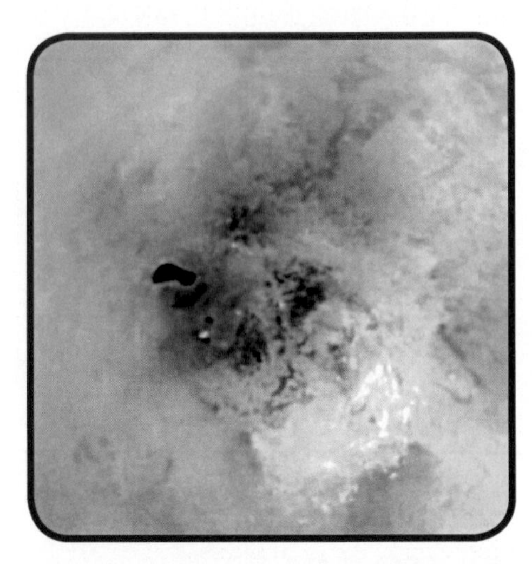

Figure 38.2b *This image was taken by the Cassini-Huygens spacecraft in 2005. It shows the surface of Titan, Saturn's largest moon.*

gravitational field is so strong that not even light can escape them. The Hubble telescope has also shown that Polaris (the North Star) is actually a cluster of three stars. It is possible to make out two of the stars with an ordinary telescope, but Hubble has spotted a third star that is so close that it has never been seen as a separate star before.

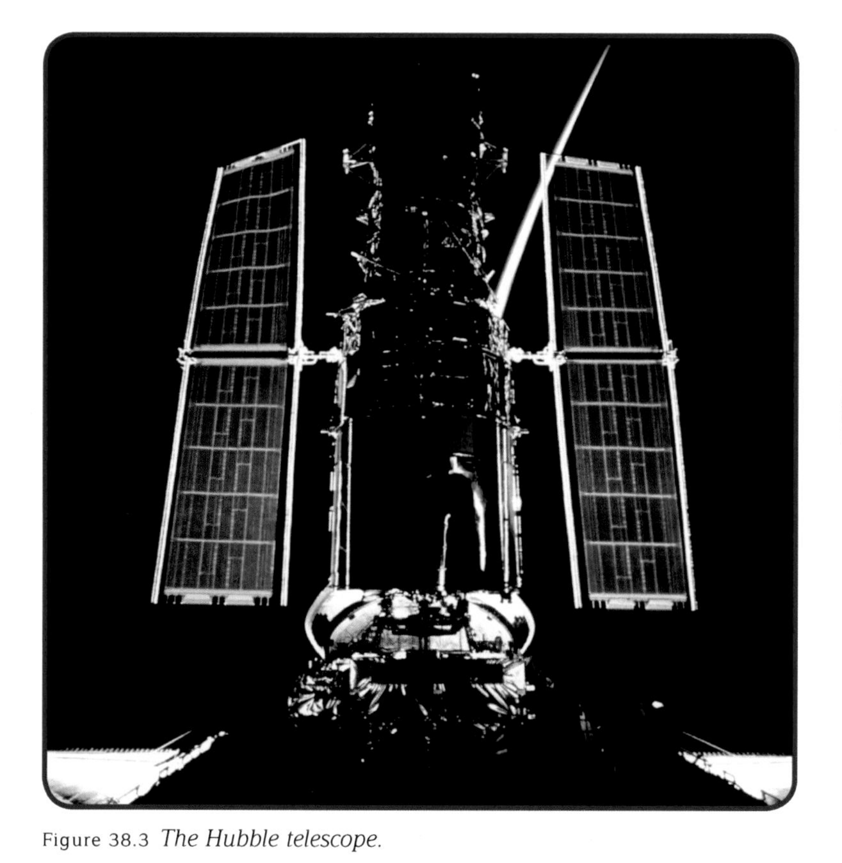

Figure 38.3 *The Hubble telescope.*

Science Extra

While the Hubble telescope mainly looks at visible and UV wavelengths of the electromagnetic spectrum, the James Webb Space Telescope (JWST), which is scheduled for launching in 2013, will detect infrared wavelengths. It will orbit Earth 2.4 million kilometres away, and because it will have a larger mirror than the Hubble, it will be able to look further into space and time.

ACTIVITY

See http://hubblesite.org for a gallery of Hubble images and details of the Hubble telescope discoveries. Find out as much as you can about the Hubble telescope and answer the following questions.

1 How big are the telescope and its mirror?
2 How long does it take to orbit Earth?
3 What are some of the important discoveries made with the Hubble telescope?

Assessment

Write a one-page essay on what you think is the most interesting Hubble telescope discovery. Or research and write a short article on the James Webb Space Telescope (see www.jwst.nasa.gov).

39

The Sun – a solar nuclear reactor

Keywords

helium, hydrogen, nuclear energy, nuclear fission energy, nuclear fusion energy, nucleus, proton, solar flare, star, supernova

Did you know?

A grain of sand heated to the same temperature as the core of the Sun would set fire to everything within a 100 km radius of it.

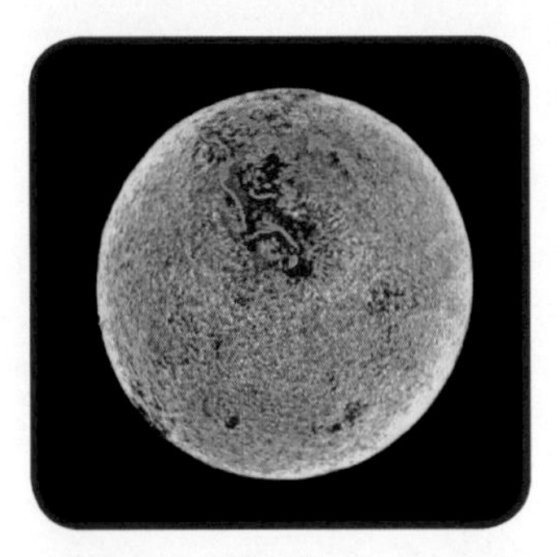

Figure 39.1 *The dark sunspots (shown by the green dots) found on the surface of the Sun.*

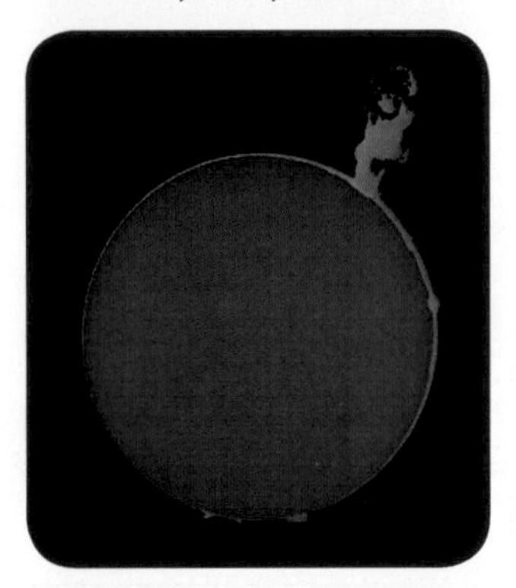

Figure 39.2 *A solar flare fires into space.*

Our star, the Sun

Compared to most other **stars**, the Sun is bright and hot. Here are some facts about the Sun:

- It is 300 000 times bigger than Earth.
- The temperature at its surface is 6 000 °C.
- The temperature at its centre is 14 million °C.
- It is 4.6 billion years old.
- Its life expectancy is approximately another 5 billion years.

Telescopic images have shown us dark sunspots that are 2 000 °C cooler than the rest of the Sun's surface, and that are accompanied by sudden and violent explosions, called **solar flares**. These activities are caused by the Sun's strong magnetic field. The Sun has a North Pole and a South Pole, and the direction of these switch about every eleven years (i.e. the North Pole becomes the South Pole and vice versa).

Solar flares can disrupt radio waves between satellites and Earth, interfering with communication systems. They can also cause electric currents in power lines, resulting in power surges and power grid overloads.

ACTIVITY

The Aztecs worshipped a Sun god, and according to Greek mythology, Helios and Apollo were gods of the Sun. As a class, collect and write up some cultural stories about the Sun.

The composition of the Sun

Anaxagoras, a Greek philosopher in 400 BCE, proposed that the Sun was made up of hot iron, and the Victorians of the 1800s believed it was made of burning coal. The table below lists the eight most common elements in the Sun. On the basis of this, can you suggest which element 'fuels' the Sun?

Element	Hydrogen	Helium	Oxygen	Carbon	Nitrogen	Silicon	Magnesium	Neon
% Composition	91.2	8.7	0.078	0.043	0.0088	0.0045	0.0038	0.0035

The Sun produces energy by nuclear fusion

As you have seen, the Sun is mostly made up of **hydrogen**. Hydrogen is the simplest element since its **nucleus** consists of a single **proton**. Protons are positively charged and normally repel each other. But in the immense heat of the Sun, four hydrogen nuclei (^{1}H) fuse or join together to produce **helium** (^{4}He). As the helium nucleus is lighter than the sum of the four nuclei, the spare mass is converted into energy (Einstein's equation $E = mc^2$ applies here, where E is energy, m is mass and c is the speed of light).

The energy that comes from the nucleus of the atom is called **nuclear energy**, and because the hydrogen nuclei have fused together, we call this **nuclear fusion energy**. For every 1 kg of hydrogen converted, 60 kJ of energy is released, and the Sun contains enough hydrogen to generate energy for another 5 billion years.

Science Extra

The nuclear fusion processes in the Sun have produced all of the elements on Earth. In the Sun, the nuclei of lighter elements join together to produce elements with larger nuclei. This is not an easy process and explains why heavier elements tend to be less common than the smaller ones.

Human-made nuclear power is nuclear fission

Ever since the first nuclear bomb was dropped in 1945, humans have been using nuclear power. This nuclear reaction is different from the nuclear reaction of the Sun. Some large elements such as uranium have nuclei that are so big they become unstable. These nuclei then split up and in the process release energy. We call this energy **nuclear fission energy** because it comes from the fission (splitting) of nuclei. Nuclear fission is controversial because the products of the reaction are radioactive and potentially deadly.

Figure 39.3 *The Sirius binary star system, showing Sirius A, the brightest star in the sky, and Sirius B (the small dot at the lower left), a white dwarf star.*

Is human-made nuclear fusion possible?

Nuclear fusion not only produces enormous amounts of energy, but, unlike nuclear fission, it produces 'clean' radiation-free energy. For this reason, a group of European scientists working on the JET project (the world's largest nuclear fusion research facility) are attempting to create nuclear fusion on Earth, imitating the production of helium from hydrogen by the Sun. The temperatures needed for this are so hot that any container would melt instantly. The superheated hydrogen nuclei are therefore kept inside a strong magnetic field.

The life cycle of a star

A star begins life as a cloud of gas. As it ages, its core dies and the outer layer swells. With the drop in temperature from white-hot to red-hot, it becomes what is known as a 'red giant'. Later, when it loses its outer layer, it shrinks in size and becomes what is known as a 'white dwarf'. Those stars that are large enough end in **supernova** explosions.

Figure 39.4 *Betelgeuse is a massive red supergiant star. Astronomers believe it is near to the end of its life and will soon explode into a supernova.*

Assessment

Explain the difference between nuclear fission and nuclear fusion.

40

Solar energy – getting our energy from the Sun

Keywords
solar energy

Did you know?

The amount of solar energy that reaches Earth every minute is more than the amount of energy the world produces by fossil fuel consumption every year.

The equator gets more solar energy than the poles

The Sun's radiant energy is released into space, and some of this strikes or reaches the Earth, having travelled a distance of 150 000 000 km. The Earth's atmosphere filters out most of the Sun's harmful radiation (UV light) and the Sun's energy reaches Earth in the form of visible light and infrared radiation. As Earth is round, more **solar energy** reaches the equator and the tropics than the polar regions. See, in the diagram alongside, how a beam of sunlight at the equator is at a right angle to the surface of Earth and covers an area the same width as the beam, while a beam of sunlight at the poles reaches the surface of Earth at more of an angle, and is spread over a wider area, i.e. it is less concentrated.

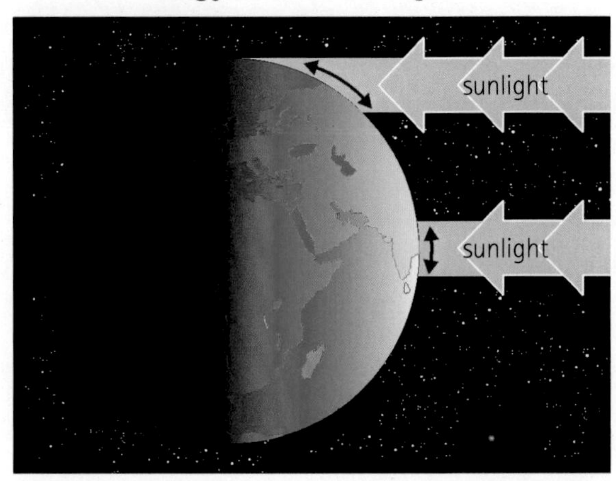

Figure 40.1 *Why solar energy is more concentrated at the equator than at the poles.*

The amount of solar energy changes with the seasons

Because of the 23.5° tilt of Earth's axis, the amount of solar heating of the polar regions varies greatly with the seasons during the year. When the North Pole is tilted towards the Sun, it gets sunlight for 24 hours of the day. Six months later, when it is on the opposite side of its orbit around the Sun, it is in darkness for 24 hours because the Sun stays below the horizon.

During their summers, the poles receive almost as much solar energy as the tropics, but in winter, they receive no solar energy at all. It is the variation in solar heating of Earth and its atmosphere that causes the climate and the seasons.

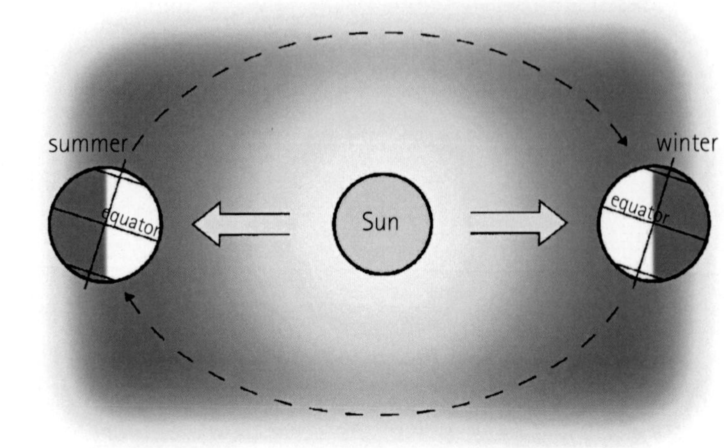

summer
equator
Sun
winter
equator

Figure 40.2 *The North Pole gets 24-hour sunlight in summer.*

Making use of solar energy

Sunlight is Earth's natural source of energy. It warms up Earth and enables green plants to photosynthesise. Some of the Sun's energy is stored in plant foods, wood and fossil fuels. The rest is lost as heat sent back into space. Given the advantages of solar energy (see the table below), energy from the Sun can be used as an alternative source of power for generating electricity and heat.

Advantages of solar energy	Disadvantages of solar energy
plentiful (in many parts of the world)	only available during the daytime and on non-cloudy days
renewable	initial installation costs are high
pollution-free	low power density, i.e. relatively large surface areas are required
low operation and maintenance costs	

Science Extra

Leaves are like natural solar panels – they take the energy from the Sun and store this as chemical energy in the form of starch. Nevertheless, photosynthetic efficiency is low. Plants capture only about 1% of the light that falls on them, or only 0.1% of the solar energy that reaches the Earth.

Solar heating systems can be used to heat household water and swimming pools. The water stored in tanks or circulating in pipes is heated by the Sun. Solar cells (photovoltaic cells) can generate electricity directly from sunlight. This is especially effective for low-power devices such as watches and calculators. Technology has also made it possible to produce electric power from solar energy.

Figure 40.3 *A domestic solar heating panel (above), and a solar power collector covered with square photovoltaic cells (below).*

ACTIVITIES

1. The map below shows how the distribution of solar energy varies according to the latitude. Explain why Trinidad and Tobago receives more solar energy than Canada.
2. Explain why the solar panel in Figure 40.3 is covered with photovoltaic cells.

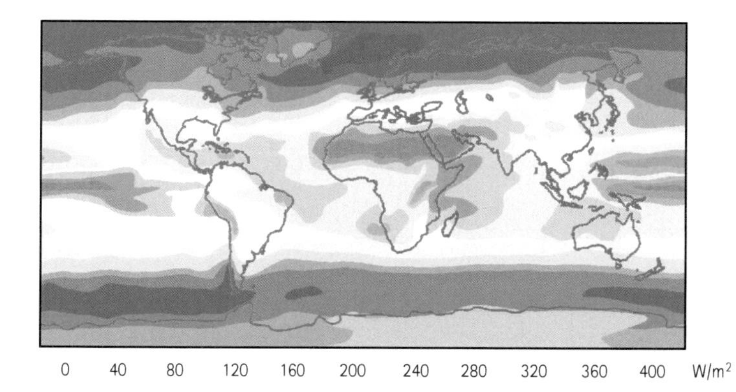

Figure 40.4 *Map showing the concentration of solar energy in the equatorial band.*

Assessment

Find out more about solar energy technology and design a solar food cooker or a solar dryer for drying sorrel or coconut.

41

The structure of the solar system

Keywords

asteroid, comet, Kuiper belt, meteor, meteorite, meteoroid, planet, small solar system body (SSSB), solar system

Did you know?

Originally, the solar system was the only known example of planets orbiting a star, but now other planets beyond our solar system have been discovered.

Our solar system

Our **solar system** consists of the Sun, eight **planets** (our own Earth and its seven neighbours), their moons, three dwarf planets with their four moons, and thousands of small bodies that include **asteroids** and **comets**. All of these circle the Sun. In the picture below, the sizes of the planets are drawn to scale, but not the distances between them. (You will look at the features of each of the eight planets in the following unit.)

Figure 41.1 *The planets of our solar system.*

The asteroid belt and meteoroids

The asteroid belt is situated mostly between Mars and Jupiter, and consists of thousands of minor 'planets' and rock-like bodies, which are possibly fragments of a shattered planet. Ceres is the largest of these and is bigger than Pluto.

If all the asteroids were lumped together, they would make an object smaller than the size of our Moon. Any asteroid that is smaller than 10 m in diameter and contains specks of rock (some of which are less than a millimetre in diameter) is called a **meteoroid**. They are normally burnt up if they pass through Earth's atmosphere, visible to us as shooting stars (**meteors**). If they land on Earth, they are called **meteorites**.

Science Extra

Where does our solar system end and the rest of space begin? At a distance three times greater than Pluto's orbit, the solar wind meets the opposing winds from other stars, forming a heliosheath, which blazes like a comet's tail. At the edge of this heliosheath, the solar wind ends, but the pull of the Sun's gravity extends further in space.

The Kuiper belt

Beyond the planet Neptune is the **Kuiper belt**. It is similar to the asteroid belt, but much broader and occupied by ice bodies. Pluto is the largest object in the Kuiper belt.

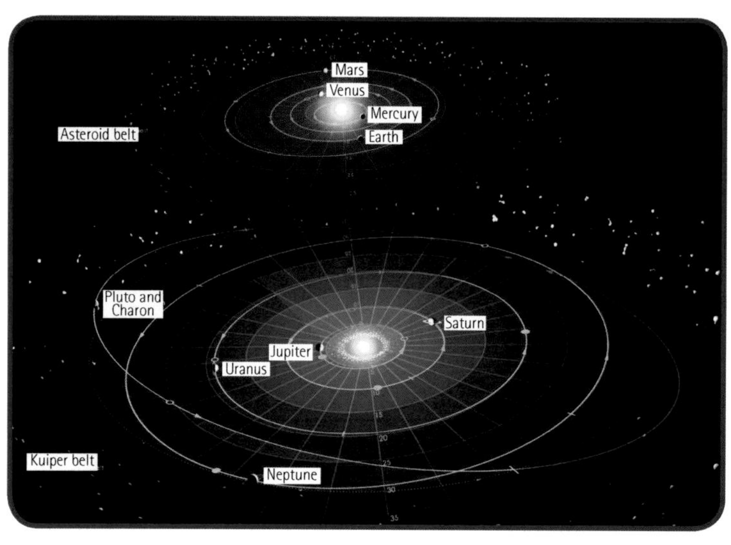

Figure 41.2 *The Kuiper belt in the solar system.*

Comets

Comets are balls of ice, and also gas and dust. Some of them, such as the famous Halley's Comet (which passes near to Earth every 76 years, the last time in 1988), are thought to come from the Kuiper belt. Comets have very unusual orbits around the Sun. Some, such as Halley's Comet, take less than 200 years to complete their orbit. Others, such as Hale-Bopp, follow paths in space that take thousands of years to complete. As a comet nears the Sun, its icy surface boils off, and it blazes a bright tail behind it. You will learn more about orbits in Unit 43.

How old will you be when Halley's Comet passes by Earth again?

Figure 41.3 *Halley's Comet, which orbits the Sun every 76 years.*

Pluto and the dwarf planets

In August 2006 the International Astronomical Union (IAU) met to come up with a definition of a planet. To be called a planet, an object must (a) orbit the Sun, (b) be round in shape, and (c) have cleared the neighbourhood around its orbit. Pluto does not fulfil the last part of this definition. It has been demoted as a planet, and is now classified as a dwarf planet.

The IAU has divided the objects in the solar system, apart from the Sun and the natural satellites or moons, into three groups: the planets, dwarf planets, and **small solar system bodies** (SSSBs). SSSBs include asteroids, the Kuiper belt objects and comets. Pluto, the asteroid Ceres, and Eris, a scattered disc object (scattered disc objects overlap the Kuiper belt), are three dwarf planets of our solar system. Charon, one of the three moons of Pluto, may be reclassified as a dwarf planet itself. Other dwarf planet candidates include the larger asteroids and objects in the Kuiper belt such as Sedna and Quaoar (pronounced 'kwa-whar').

Figure 41.4 *The cratered rock and ice surface of Pluto (right) and its moon, Charon (left).*

Assessment

Look at the picture of the solar system on page 86. Identify the asteroid belt and each of the eight planets. (Use Unit 42, page 88 to help you if necessary.)

Figure 41.5 *Quaoar (above) and Sedna (left) are two of the largest objects in the Kuiper belt.*

The eight planets

Keywords

Earth, Jupiter, Mars, Mercury, Neptune, Saturn, Uranus, Venus

Did you know?

Except for Earth, the planets are named after Roman gods and goddesses. Mercury is named after the winged messenger of the gods because it is the fastest planet to circle the Sun, taking only 88 days. Neptune is named after the god of the sea because of its bright blue colour.

Mercury, Venus, Earth, Mars, Jupiter, Saturn, Uranus, Neptune

The eight planets of our solar system (shown on page 86) make up a very diverse family. They range from the tiny and hot **Mercury** to the enormous gas giant **Jupiter**, with its 28 moons and swirling storm cloud, known as the Great Red Spot. They range from **Mars**, which is red because of its rusted iron surface, to **Uranus**, which is blue with a surface of liquid methane. Then there's beautiful **Saturn** with its bright rings made up of small lumps of ice and its many moons including Dione, Enceladus, Rhea and Titan.

The first four planets – Mercury, **Venus**, **Earth** and Mars – are known as the terrestrial planets. They all have a solid outer crust of rock similar to the Earth. The next four planets – Jupiter, Saturn, Uranus and **Neptune** – are called the gas giants. They are much larger than the terrestrial planets, and have gaseous outer layers that probably surround a solid core.

The table below gives some of the main features of each planet. Mercury, Venus, Mars and Jupiter are all visible to the naked eye. Uranus can be viewed with binoculars and Neptune through a small telescope.

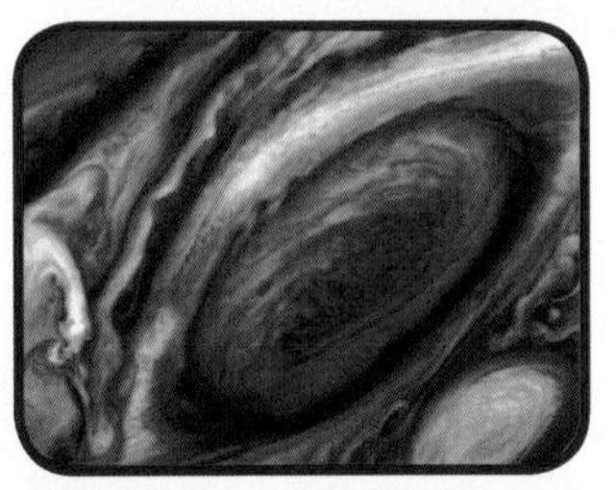

Figure 42.1 *Jupiter's Great Red Spot.*

Figure 42.2 *Saturn's rings.*

Figure 42.3 *Planet Earth.*

Planet	Mercury	Venus	Earth	Mars	Jupiter	Saturn	Uranus	Neptune
Distance from Sun (Earth is 1)	0.3871	0.7233	1	1.524	5.203	9.539	19.19	30.06
Mass (Earth is 1)	0.06	0.82	1	0.11	317.89	95.18	14.53	17.14
Radius (km)	2 439	6 052	6 378	3 397	71 490	60 268	25 559	25 269
No. of moons	0	0	1	2	63	56	27	13
Density (g/cm^3)	5.43	5.25	5.52	3.95	1.33	0.69	1.29	1.64
Day length (hours)	1 408	5 832	23.93	24.62	9.92	10.66	17.24	16.11
Length of year (Earth years)	0.24	0.62	1	1.88	11.86	29.46	84.01	164.79
Mean surface temperature (°C)	167	657	14	−55	−153	−185	−214	−225
Atmosphere	almost none	CO_2, N_2	N_2, O_2	CO_2, N_2	H_2, He	H_2, He	H_2, He	H_2, He

ACTIVITIES

1 Which planet am I?

a Solve these riddles:

- I have no moons and am similar in size to Earth.
- I have a day that is 24 hours long and I am red with rust.
- I have a Great Red Spot.
- I am the only planet with oxygen in my atmosphere.
- My day is shorter than an Earth day, and my density is less than that of Uranus.
- I am the coldest planet and the furthest from the Sun.
- I am not much bigger than Earth's moon and I am hot and hostile.
- Of all the planets, I rotate the most slowly. My 'day' is longer than my 'year'.

b Make up some more of these questions yourself and test a friend. Each planet is unique and there are lots of interesting things to learn about each one. Venus, for example, is often referred to as the 'Morning star' or 'Evening star'. Find out more, either individually or in pairs, about a planet that appeals to you and then present the information as a poster or an oral presentation.

2 The diagram on page 86 shows the relative size of the planets, but not how far apart they are from each other. The table on page 88 gives the distance of each of the planets from the Sun in Astronomical Units (AU). One AU is defined as 150 000 000 km, the distance from Earth to the Sun.

a To get a sense of the scale of our solar system, pace out the distances of the planets from the Sun on the sports field. OR using a scale of 1 cm = 1 AU, mark out the distances of the planets from the Sun on a piece of paper.

b What do you notice about the distribution of the planets? Write up in your own words a brief description of how the planets are positioned in the solar system.

3 'My Very Educated Mother Just Served Up Nine Pizzas' was a common mnemonic for remembering the order of the nine planets when Pluto was still classified as a planet. Have a competition in class to see who can come up with the best mnemonic for the order of the eight planets.

Science Extra

The solar wind, a blast of charged particles emitted by the Sun, has removed most of Mercury's protective atmosphere. Venus is the hottest planet of the solar system, probably because of its very thick atmosphere, which traps in the heat. The Venus Express Satellite probe has been analysing Venus' atmosphere, which is made up of 96.5% carbon dioxide and 3.5% nitrogen.

Assessment

Make up a calypso for the eight planets, describing the uniqueness and beauty of each.

The motion of the planets

Keywords

elliptical, gravity, orbit, rotate

Did you know?

The longest orbit in our solar system is now that of Sedna, the dwarf planet discovered in 2003. Sedna takes 12 050 years to complete its orbit, i.e. a Sedna year is 12 050 Earth years.

The planets orbit the Sun

The name *planet* means *wanderer*. All the planets in our solar system circle around the Sun, and the path that a planet follows is called its **orbit**. The diagram below shows the orbits of the planets of our solar system. Because the orbits are all **elliptical** shapes (squashed circles), each planet's distance from the Sun varies slightly during the year.

Figure 43.1 *The orbits of the planets of our solar system.*

All the planets orbit the Sun in the same direction (anti-clockwise) and in the same plane, meaning that their paths are all on the same 'level' relative to the Sun. The dwarf planet Pluto has an (elliptical) orbit tilted at 17°.

How long is an orbit?

We call the time it takes a planet to complete one orbit of the Sun a year. Earth's year is $365\frac{1}{4}$ Earth days. The other planets of the solar system take different lengths of time to orbit the Sun. The further a planet is from the Sun, the greater the distance it must travel, and the slower its speed. Those planets closest to the Sun (where the Sun's gravity is the strongest) circle the Sun the fastest. The table below shows the length of a year for each of the planets.

Planet	Mercury	Venus	Earth	Mars	Jupiter	Saturn	Uranus	Neptune
Length of year	88 Earth days	225 Earth days	$365\frac{1}{4}$ Earth days	687 Earth days	12 Earth years	29 Earth years	84 Earth years	165 Earth years

ACTIVITIES

1. Demonstrate an orbit.
   a. Tie a tennis ball or a small stone to a piece of string.
   b. Swing it in a circular motion above your head.
   c. Observe how, although you are constantly pulling the ball or stone towards you, it moves sideways and therefore in a circular motion.
2. Use the tables on pages 88 and 90 to answer these questions.
   a. Roughly how does the time a planet takes to complete an orbit depend on its distance from the Sun?
   b. How many times does Earth orbit the Sun in one Jupiter year?
   c. If you lived on Jupiter, how long would you say a Uranus year was in Jupiter years?

Planets are held in their orbits by gravity

The solar system formed out of a spinning cloud of gas and dust. **Gravity** is the main force governing the universe and our solar system, and it is because of gravity that planets are round in shape and that they orbit the Sun. As the Sun is the biggest object in our solar system, it has a strong gravitational pull on objects in its vicinity, and the planets of our solar system are trapped in its gravitational field. But what prevents the planets from being pulled into the Sun? At the same time as a planet is being pulled by the Sun, it is moving at right angles to the Sun's pull on it.

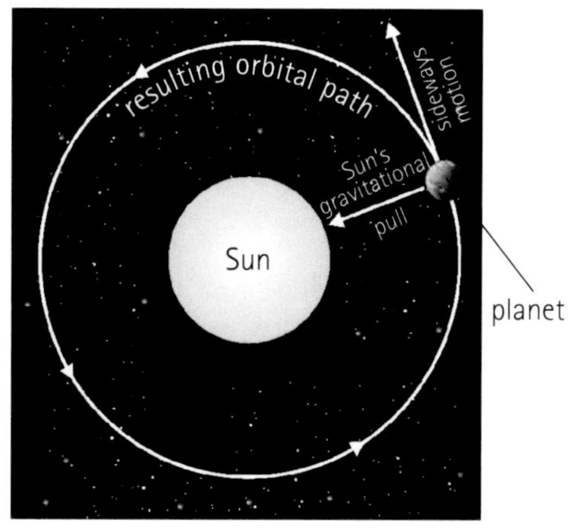

Figure 43.2 *As well as being pulled towards the Sun, a planet moves sideways.*

Day and night – the planets spin as they orbit the Sun

As the planets circle around the Sun, they also spin or **rotate**. Earth spins around its axis (which is tilted at an angle of 23.5°) every 24 hours.
All of the planets, except for Venus and Uranus, rotate anti-clockwise. This means that on Venus, unlike on Earth, the Sun rises in the west, and sets in the east.

ACTIVITY

1. Look at the table of the planets on page 88. Work out the relationship between the speed at which a planet spins and its size (excluding Venus). Do this by:
   - ordering the planets from smallest to largest, according to mass
   - then ordering the planets from the longest day to the shortest day.
2. Look at the pattern and then choose the correct option for the following statement: The biggest planets spin the slowest/fastest.

Science Extra

If you were on Venus, you wouldn't see the Sun rise very often – only every 243 Earth days – because it rotates very slowly compared to the other planets. Uranus' axis of rotation is at 98° and Venus' at almost 180°, which means that Uranus is rotating on its side, and Venus rotating anti-clockwise.

Assessment

Describe the difference between the orbiting and spinning motion of a planet.

44

The Moon – our travelling companion

Keywords
crescent, gibbous, moon, tide, waning, waxing

Did you know?

Isaac Newton was the first person to give a scientific explanation of the tides.

The Moon is our nearest neighbour

In the same way that planets circle the Sun, **moons** orbit their home planets. The Moon is our own natural satellite or travelling companion and orbits the Earth, taking about 28 days to complete each orbit. In other words, it takes 29 days from one full moon to the next full moon. The Moon, as our nearest celestial neighbour, is the largest and brightest object in the sky. But why does it appear to change its shape as it orbits Earth?

Figure 44.1 *The Moon orbits Earth (the blue path), while Earth orbits the Sun (the red path).*

The phases of the Moon

One side or face of the Moon is always lit up by the Sun, but we see different portions of this lit-up face, depending on the position of the moon in its orbit. Only when the Moon is directly opposite the Sun do we see the whole of the illuminated disk, and we call this a full moon. When the Moon is between Earth and the Sun, we cannot see any of the lit-up side, and we call this a new moon. There are eight distinct phases of the Moon, beginning with new moon.

Figure 44.2 *The phases of the Moon during a lunar month. The full moon is at the centre.*

ACTIVITY

Use Figures 44.1 and 44.2 to answer the following questions.

1 How long between new moon and full moon? When the moon is full, how far is it in its orbit around Earth?

2 Choose the correct option each time: **Waxing** means getting bigger/ smaller and **waning** means getting bigger/smaller. A **crescent** moon is less/more than a quarter moon, and a **gibbous** moon is less/more than a quarter moon. (*Gibbous* comes from the Latin word meaning *humped*.)

3 Explain why at new moon, the Moon and the Sun rise and set at about the same time, while at full moon, the Moon rises as the Sun sets.

The Moon and the tides

The Moon circles Earth because of the Earth's strong gravitational pull on it. But the Moon also pulls on Earth and its oceans, and its gravity is strong enough to cause the **tides**. The pull of the Moon makes the sea bulge on the side of Earth that is facing it, as well as the opposite side. These bulges form the high tides, and as Earth rotates, the bulges move, and so the tide at a particular place changes from high to low and back again about every 12 hours. The time interval between one high tide and another is 12 hours and 26 minutes because, at the same time as Earth is rotating, the Moon is orbiting the Earth. This means that, at the same time each day, the Moon is not where you saw it the day before, but is still 'catching up'.

The Sun also pulls on the Earth, affecting the tides. At full moon and new moon, when the Sun, Moon and Earth are lined up, the combined pull of the Sun and the Moon creates the biggest effect, causing spring tides (i.e. high highs and low lows). At the quarter and three-quarter moon phases, the Sun and the Moon are pulling at right angles to each other, creating the weakest effect, i.e. neap tides.

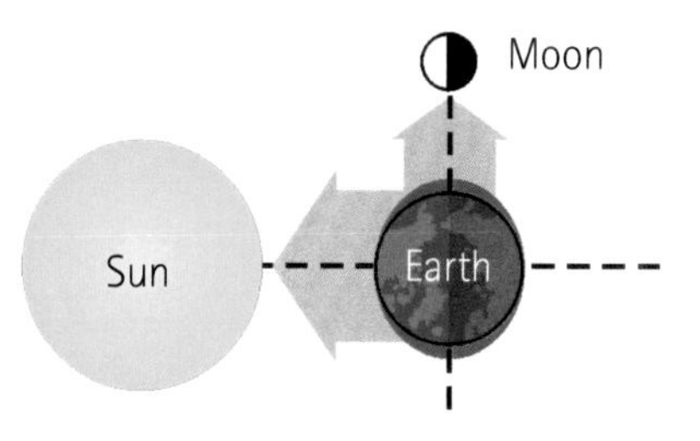

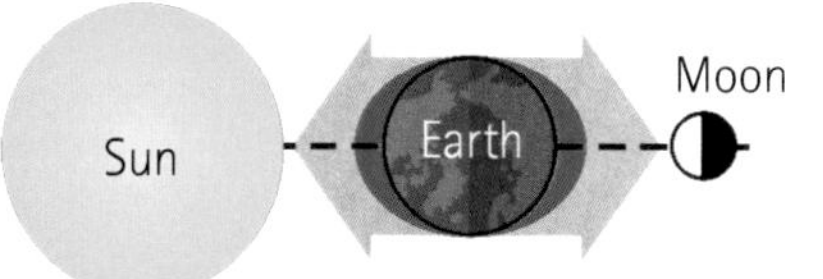

Figure 44.3 *How the tides form.*

INVESTIGATION Moon and tide calendar

Observe the phases of the Moon for a month-long period, keeping a record. Present this in a report with a tide table and/or tide graph for the same period. You can access a tide table and tide graph for Tobago on www.surf-forecast.com/breaks/BeachBreak.tide.shtml. Describe the relationship that you see between the phases of the Moon and the high and low tides.

The surface of the Moon

The surface of the Moon is pockmarked by craters that have been formed by meteorites. You can take a closer look with a pair of binoculars or a hand-held telescope. Because the Moon has no atmosphere, it is unprotected and has been easily scarred by meteorites that have struck its surface and exploded on impact.

Figure 44.4 *The Moon has mountains and valleys. This picture shows Crater Timocharis on the Moon.*

The origins of the Moon

There have been several theories about where the Moon came from. The most generally accepted theory is that the Moon is debris from a planetoid collision with Earth long ago.

Science Extra

Have you ever noticed that although it changes shape, the surface of the Moon always has the same appearance? The same side of the Moon always faces us because the Moon rotates on its axis at exactly the same rate as it orbits Earth. We call the side of the Moon that we never see, the 'far side', and only astronauts have seen it.

Assessment

Explain the phases of the Moon to a friend who is not in your class, using a diagram.

Eclipses of the Sun and the Moon

Keywords

annular eclipse, corona, lunar eclipse, solar eclipse

Did you know?

There are at least two solar eclipses (partial) every year somewhere on Earth. A total solar eclipse occurs somewhere on Earth about every one and a half years, but is only likely to be visible in the same place every 370 years.

An eclipse of the Sun

There was a 90% eclipse of the Sun in Trinidad and Tobago in 2000. Anyone who has seen a **total solar eclipse** – which takes place when the Moon passes directly between the Sun and the Earth, blocking out our view of the Sun – cannot fail to be impressed. The image of the Sun being 'eaten' by the Moon (and the accompanying darkness and temperature drop during the daytime) is so strange that people have recorded solar eclipse events throughout history. As you have learnt in the previous unit, the Moon is between the Sun and the Earth only during the new moon phase.

Figure 45.1 *An eclipse of the Sun is called a solar eclipse.*

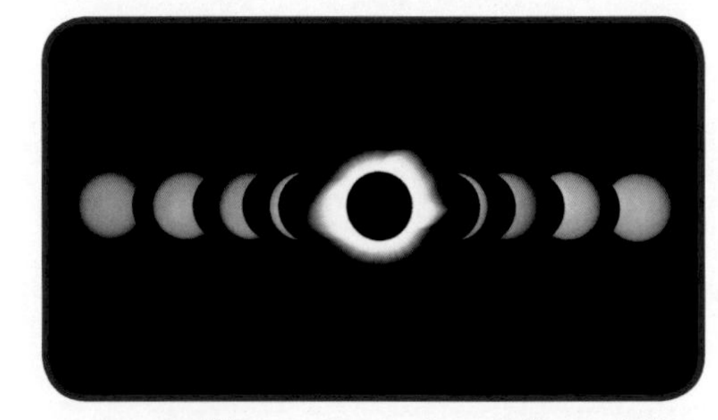

Figure 45.2 *The sequence of an eclipse of the Sun.*

The pictures in Figure 45.2 show a sequence taken during a total solar eclipse. As the Moon passes in front of the Sun, the Sun begins to disappear from view, until it is completely blocked out by the disk of the Moon. During this time you can see a fiery **corona** of light that is just visible around the edge of the Moon. This lasts for about two minutes, and then the Sun begins to reappear.

A total eclipse takes place when the whole of the Sun is blocked out, and an **annular eclipse** occurs when the Moon appears smaller than the Sun, and so the Sun is seen behind the disk of the Moon as a bright ring. (The 'size' of the Moon's disk varies because the Moon's orbit is elliptical, i.e. at some points in its orbit, the Moon is closer to Earth than at other times.) A partial eclipse occurs when only part of the Sun is masked by the Moon.

How to view a solar eclipse safely

It is important never to look at the Sun directly while the eclipse is taking place, since staring directly at the Sun can do permanent damage to your eyes. View the eclipse through special viewing glasses or the thin foil packaging sometimes used for teabags. The safest way to view the eclipse is indirectly, by projection (see page 111).

Eclipse of the Moon

An eclipse of the Moon, called a **lunar eclipse**, takes place when the Sun, Earth and Moon line up with Earth in the middle.

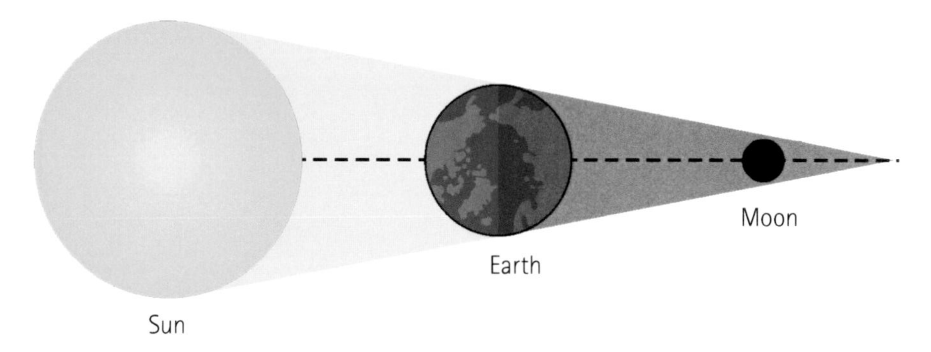

Figure 45.3 *A lunar eclipse: when the full moon is in the Earth's shadow.*

A lunar eclipse only happens during a full moon, when the Moon is on the opposite side of Earth to the Sun. As the Moon passes in front of the Earth, the Earth's shadow falls on it. You would expect the Moon to completely disappear because no light would reach it, but the atmosphere of Earth bends some light from the Sun around to the Moon, making the Moon appear red or orange in colour.

Figure 45.4 *The orange lunar eclipse.*

ACTIVITIES

1 People have always developed stories and beliefs about eclipses. The ancient Chinese explained solar eclipse events as a dragon in the sky trying to swallow the Sun. An Indian superstition is that a pregnant women exposed to the eclipse will give birth to a blind baby or one with a cleft lip. Find out about the stories associated with eclipses and discuss them in class.

2 Make the room as dark as you can and demonstrate the formation of eclipses as follows:
   - Get a friend to shine a torch as the Sun.
   - Imagine you are the Earth. Hold a ball or orange in your hand to represent the Moon.
   - To demonstrate a solar eclipse, slowly move the Moon directly between you and the Sun. Can you see how the light from the Sun seems to disappear? Create an annular eclipse by experimenting with the distance between you and the Moon.
   - To demonstrate a lunar eclipse, move the Moon into your shadow. Can you see how the light that was reflected by the Moon disappears?

Science Extra

You might think from what you have learnt about the Moon's orbit in Unit 44 that there should be an eclipse of the Sun every month during new moon. In fact, the orbit of the Moon around Earth is at a slight angle compared to the orbit of Earth around the Sun, so it does not line up very often. The same is true for lunar eclipses.

Assessment

Make a poster to show how solar and lunar eclipses are formed. Take this home and explain to your family what you have learnt.

46

Satellites for communication, observation and navigation

Keywords

geographical information satellite (GIS), geostationary orbit, global positioning system (GPS), low polar orbit, satellite

Did you know?

There have been so many satellites launched that the upper atmosphere is filling up with space junk. There are more than 23 000 items of 'junk' in space.

Figure 46.1 *A communication satellite in space.*

A satellite is any object that orbits a planet

You have already learnt about one natural **satellite** – Earth's moon – and one artificial satellite – the Hubble telescope. With developments in technology, artificial satellites are now a part of everyday life. When you watch satellite TV, you are seeing pictures that have been relayed from space. When you phone a friend in another country, you are probably talking to them via satellite. And when you read the weather forecast, the predictions have been made using weather data collected by satellites.

Communication satellites in geostationary orbits

Television and telephones make use of communication satellites. These satellites need to be in a fixed place so that the broadcast isn't disrupted. But how do they stay in one place above Earth when Earth is rotating?

The satellites are placed in what is called a **geostationary orbit**. This orbit is at a height of 36 km above the Earth's surface, so that the satellites orbit Earth at exactly the same speed as Earth spins. In this way, the reception antennae, which point in a fixed direction, maintain unbroken contact with the satellite.

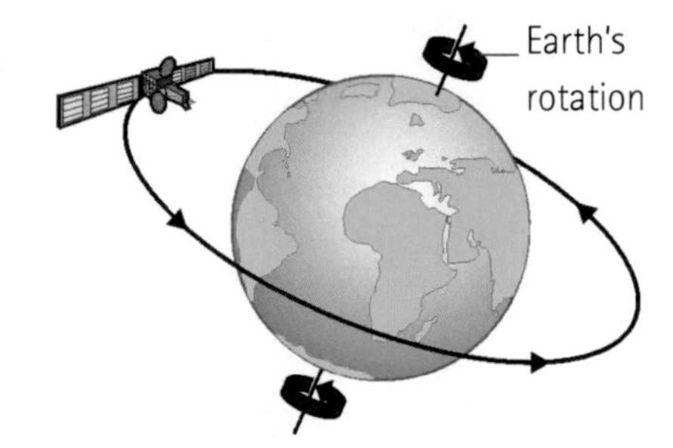

Figure 46.2 *Communication satellites stay in one place above Earth.*

Geographical information satellites in low polar orbits

By contrast, satellites that are used to collect information about the weather need to move to collect data. They also need to be nearer to the surface of Earth. These **geographical information satellites (GISs)** are placed in what is called a **low polar orbit** over the North and South poles. Their orbit is at right angles to the direction of Earth's rotation. They orbit Earth every 45 minutes, while Earth rotates underneath them. In this way, with successive orbits, they cover the whole planet.

You can easily see low polar orbit satellites in the night sky. Look out for what looks like a fast-moving star.

Figure 46.3 *Weather systems are tracked using GIS satellites. This is Hurricane Isabel in the Caribbean Sea on 15 September 2003.*

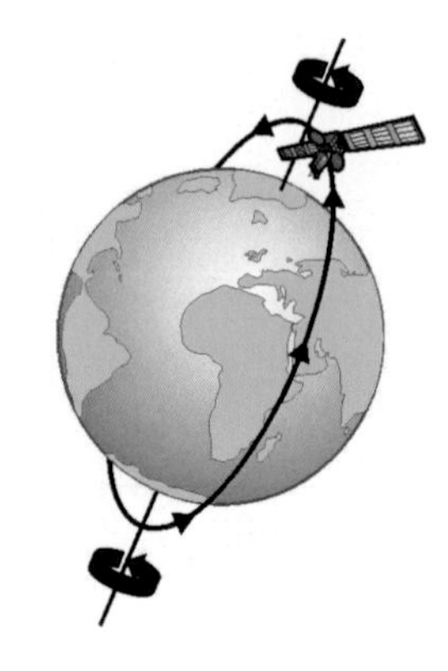

Figure 46.4 *Weather satellites follow low polar orbits.*

ACTIVITIES

Collect the weather system pictures published in the newspaper. Write an article for your school magazine about the importance of GIS satellites.

OR

Many institutions in Trinidad and Tobago (e.g. the Water and Sewerage Authority, the University of the West Indies, and the Institute of Marine Affairs) use GIS systems to map out or monitor various projects in our country. Find out one such project that uses GIS technology, and give a brief description of that project.

Global Positioning System (GPS) array

Satellites are also used for navigation. Aircraft, boats and some cars are equipped with **global positioning system (GPS)** devices. These small devices receive radio signals from a network of more than 24 satellites orbiting Earth, and use these signals to pinpoint their own position within a few metres. A GPS receiver calculates its position by measuring the distance between itself and three or more GPS satellites. It can work out not only location, but also speed and direction.

Figure 46.5 *Some cars have GPS devices in them.*

How is a satellite launched into an orbit?

Satellites are launched into space riding on a rocket. Usually the rocket is aimed straight up, the quickest path through the atmosphere. The rocket is equipped with an internal guidance system that follows a flight plan, and it heads east, because Earth rotates east and this rotation gives the rocket an extra boost. (The strength of the boost depends on the rotational velocity of Earth, which is greatest at the equator.) Once about 200 km up, the rocket's navigational system fires small rockets to turn itself into a horizontal position and the satellite is then released.

To escape Earth's gravity and fly off into space, a rocket must accelerate to more than 40 000 km/h (this is called *escape velocity*). But to place a satellite in orbit requires a balance between the pull of gravity on the satellite and the inertia of the satellite's motion (you will learn about inertia in Topic 5). At the correct orbital velocity, the pull of gravity is just enough to keep the path of the satellite curving around Earth's surface. The orbital velocity depends on the altitude or distance of the satellite from Earth. The nearer it is to Earth, the faster it must travel.

Science Extra

Satellites use renewable energy. They have large solar panels that open out once they reach their orbit. Because there is little or no atmosphere, there is plenty of sunlight in space.

Assessment

Brainstorm all the different ways that satellites influence our lives. Do you think they have influenced our lives for good? Explain your thinking.

47

Journeys into space

Keywords
astronaut, International Space Station, microgravity, weightlessness

Did you know?
The person who holds the record for the longest stay in space is Valeri Polyakov, a Russian who spent 437.7 days in space after launching on 8 January 1994.

The first trips into space

Sputnik 1 was the first human-made object to orbit Earth in 1957, and thirty days after the *Sputnik* launch, the first animal – a dog called Laika – was launched into space. After 92 days, *Sputnik* was pulled back into Earth's atmosphere by gravity and burnt up, and the satellite carrying Laika followed the same fate.

Yuri Gagarin was the first human in space, in *Vostok I* in 1961, and the first space walk was done by Aleksi Leonov four years later. Then, in 1969, after a five-day trip from Earth, *Apollo II* accomplished the first Moon landing, and Neil Armstrong became the first person to set foot on the Moon. So far, the Moon is the furthest distance in our solar system that **astronauts** have travelled.

These first ventures into space marked a 'space race' between the Soviet Union and the United States of America. In the 1980s, the USA began its Space Shuttle Program, engineering reusable spacecraft – *Columbia*, *Challenger*, *Discovery*, *Atlantis* and *Endeavour*. The *Columbia* was the first shuttle in space in 1981, but disintegrated while re-entering Earth's atmosphere on a flight in 2003, killing all seven astronauts on board. The *Challenger* suffered a similar disaster during launch in 1986. Although humans haven't ventured far, space exploration has come a long way since it began, and has changed from a competition to an international collaboration.

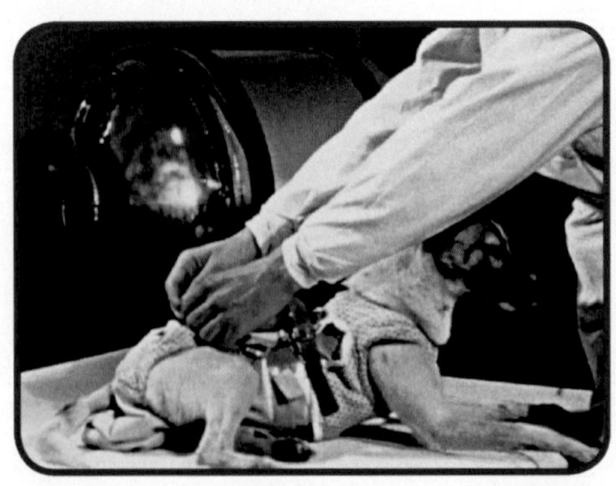
Figure 47.1 *Laika, the first animal astronaut.*

Figure 47.2 *An astronaut saluting the American flag on the Moon after Apollo 16 landed on 21 April 1972.*

Science Extra
The smallest personal computer you could buy today is more powerful than the computer that landed the first spacecraft on the Moon.

ACTIVITY

Research the key events in the history of space exploration. Make a timeline for your classroom.

The International Space Station

With precision, an overhead crane swings a 10-ton building block into position. Then, workers move in, climbing onto the structure and using hand and power tools to bolt the pieces together. This construction site is 360 km above the Earth. The construction workers are astronauts, the cranes are a new generation of space robotics and the skyscraper taking shape is the International Space Station.

(Adapted from NASA's Shuttle Press Kit)

Figure 47.3 *An astronaut at work on the construction of the International Space Station in October 2000.*

The assembly of the **International Space Station** (ISS) began in 1998 to replace the Russian Mir space station that orbited Earth from 1986 to 2001. It is a collaboration between 16 countries. The *Endeavour*, *Discovery* and *Atlantis* space shuttles have all carried component parts to the ISS, and it will take about 40 assembly flights to complete, and cost at least US $35 billion.

The ISS orbits Earth about every 90 minutes and is powered by a huge array of solar panels. It is made up of pressurised modules, containing the same mixture of air at the same atmospheric pressure as on Earth. There is always a crew of at least two astronauts on the space station, and it is being designed to house a crew of six to seven. It has already hosted visiting space tourists.

The ISS will function mainly as a laboratory in space, making it possible to find out more about the effects of gravity in a **microgravity** environment, where gravity has little or no effect.

Figure 47.4 *The International Space Station with Earth in the background.*

Weightless in space

One of the things that astronauts must cope with in space is **weightlessness** – the feeling of having no weight. Astronauts often get 'space sickness' as they adjust to the weightlessness, with headaches, nausea and vomiting that can last up to 72 hours.

A microgravity environment affects the long-term health of astronauts. Without the workings of gravity for proper exercise and proper circulation of the blood down into the legs, muscles get weaker and bone mass reduces. Astronauts can develop thin 'chicken legs' and puffy faces, and generally lose fitness. Studies have also shown that weightlessness can weaken the immune system and decrease the production of red blood cells.

In order to stay physically healthy, astronauts must spend a good part of each day exercising, but this is tricky without gravity. There is no point in 'lifting' weights in an environment where objects float. Astronauts have to strap themselves onto the treadmill or the exercise bicycle.

Figure 47.5 *Life aboard the International Space Station in a microgravity environment.*

Assessment

List some of the health problems that you think might be associated with staying in a weightless environment. Why might astronauts have trouble walking when they return from a space mission?

48

Mission to Mars

Keywords
colony

Did you know?

Mars has the highest mountain of all the planets and the biggest canyon, which is 4 000 km in length and 7 km deep in some parts. By comparison, the Grand Canyon in the USA is about 450 km long and its maximum depth is 2 km.

Figure 48.1 *A Mars Rover on the surface of Mars.*

Exploring Mars

In 2004 NASA landed two rover probes, *Spirit* and *Opportunity*, on Mars. Since then, these small, robotic, solar-powered vehicles have been exploring the planet (travelling, digging, grinding rock, photographing) and sending back pictures of what they see – clouds, whirlwinds, a meteorite, and hills and craters. Despite a few hiccups such as flash memory computer failure and getting stuck in ditches and sand dunes, the mission lifespan of these two robots has already been ten times longer than expected.

In addition, there are four spacecraft orbiting the planet – the *Mars Global Surveyor*, the *Mars Odyssey*, the *Mars Express* and the *Mars Reconnaissance Orbiter*. One of the goals of the *Odyssey* mission was to measure the radiation levels on Mars, which are 2–3 times higher than on Earth. This is because Mars has less than 1% of Earth's protective atmosphere and no magnetic shielding from solar flares and cosmic radiation (which can cause cancer and damage the central nervous system).

The images from *Odyssey* have been used to help plot the landing sites for the rovers and to guide the rovers to rocky terrain; this is because rocks are more interesting to study than dust. Most of the rover images have been relayed to Earth via *Odyssey*, which completes a polar orbit around Mars every two hours and passes over both the rovers at least twice a day.

Figure 48.2 *An image of the surface of Mars taken by a Mars Rover camera.*

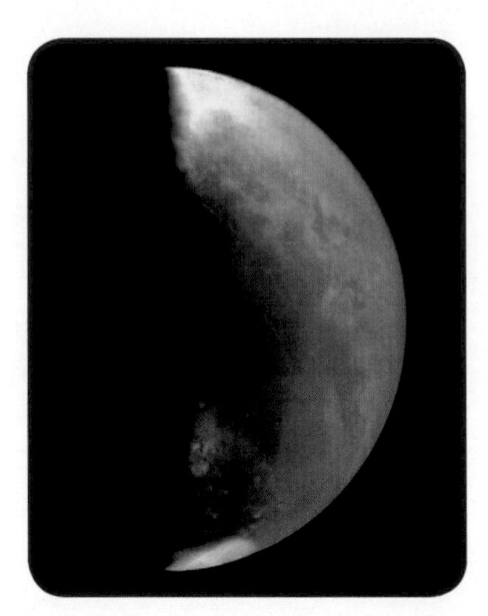

Figure 48.3 *A photograph of half of Mars, showing the polar ice caps.*

Living on Mars?

Although Mars is not our nearest neighbour (Venus is), it is the planet most similar to our own and therefore a good choice of planet on which to set up a human **colony**. A Martian day is about 24 hours long (it takes the same time as Earth to complete one rotation). And because the tilt of its axis is almost the same as Earth's, its seasons are similar, although twice as long. Its temperatures range from –140 °C in the winter to 20 °C in the summer.

Figure 48.4 *The Mars landscape at sunset.*

From the dents and grooves on Mars' surface, it appears that there were rivers that flowed long ago. If there was liquid water on Mars long ago, there could also have been life, and the frozen polar ice caps of carbon dioxide might also contain water that can be accessed to support human life. Mars has a very thin atmosphere, though, and therefore almost no insulation. Because of its small size, its gravity is only one third of the Earth's, and most of its atmosphere was probably lost to space as the planet heated up and gravity was not strong enough to hold it in place.

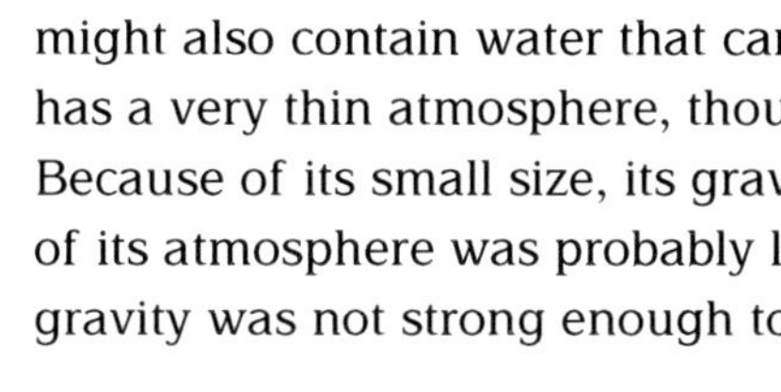

ACTIVITY

Only the far distant future will tell us whether humans living on Mars is science fiction or reality. But if humans are going to live there one day, we need to start planning for it now. As a group, you are responsible for coming up with a plan to establish a base on Mars, which you will present to the rest of the class. Make use of the information given in this unit and compare Earth and Mars in the tables on pages 88 and 90. You can also do research on the Internet to find out more – 'Colonization of Mars' on the Wikipedia site: http://en.wikipedia.org/wiki/Colonization_of_Mars is a good place to start. The information in the margin and the questions below should help you develop your ideas.

- How many people would you take to colonise Mars, and who would you choose?
- What supplies would you take, considering cargo space is limited?
- Where on Mars would you set up your base?
- What sort of structure would you live in?
- How would you protect yourself against radiation?
- How would you get oxygen to breathe?

Assessment

Is it right that humans should consider setting up a base on another planet? Is it simply a solution to over-population and destruction of our own planet? Or is it necessary to ensure the survival of the human race when the Sun burns itself out in billions of years' time? Write a one-page essay on what you think.

OR

Present a proposal to NASA for the exploration of a particular planet. Give full reasons for choosing this particular planet.

Review topic 3

Compared to the universe that is populated with billions of stars, compared to our galaxy the Milky Way, and even compared to the whole of the solar system, our place in space (on the planet Earth) is very small. Earth is one of the eight planets that orbits a star, the Sun, which generates heat and light (solar energy) by nuclear fusion. The other planets in our solar system are the three terrestrial planets and our closest neighbours – Mercury, Venus and Mars – and the four gas giants much further away – Jupiter, Saturn, Uranus and Neptune. Many of these planets have their own moons which circle around them, and Earth has its own single Moon, whose gravity is responsible for the sea tides. Occasionally, when the Moon passes between Earth and the Sun, Earth experiences an eclipse of the Sun. The other bodies in the solar system include dwarf planets (of which Pluto is one), asteroids and comets.

Artificial satellites are manufactured objects that orbit Earth or the other planets of the solar system, and are used for communication, navigation and exploration or observation. The Hubble telescope is one such satellite that orbits 600 km above Earth, looking far into space beyond Earth's distorting atmosphere. The International Space Station is another giant satellite that circles Earth and serves as a research station for astronauts in a space environment, where the Earth's gravity has little effect.

So far, the Moon is the furthest destination travelled to by humans in space, but the nearby planets such as Mars and Venus are being explored by satellites, and the *Spirit* and *Opportunity* rovers have been exploring the surface of Mars.

Multiple-choice questions

1 A pattern of stars is called a
   a galaxy
   b constellation
   c solar system
   d universe
2 Which is the largest planet of our solar system?
   a Mars
   b Venus
   c Earth
   d Jupiter
3 Which of the following statements about the Sun is incorrect?
   a It generates energy by nuclear fission.
   b It consists mainly of hydrogen.
   c It is the only star in our solar system.
   d It has a north pole and a south pole.

Longer questions

1 How long does it take the Sun's light to reach the Earth? (To answer this, look up the distance between the Sun and Earth, and the speed of light.)
2 a Name the eight planets of our solar system, in order from closest to furthest from the Sun.
   b How do the distances of the planets from the Sun affect their ability to sustain life?
   c Which planet of the solar system is no longer classified as a planet, and why?
   d Explain the difference between an asteroid and a comet.
3 For the motion of a planet, explain the difference between orbit and rotation.
4 If you see the crescent moon like this ☾ in the sky, is it waxing or waning?
5 Draw diagrams to show
   a how a solar eclipse is formed
   b how a lunar eclipse is formed.
6 a Draw a diagram showing how a communication satellite helps to transmit signals around Earth.
   b What is meant by GPS satellites and systems?
7 Write an essay describing how humankind has explored the solar system. Do you support the expenditure of money on these ventures?
8 Draw a scaled diagram showing the various planets in our solar system and their distances from the Sun.
9 Write a chronological account about human exploration in space. OR draw a timeline to summarise human space exploration, starting with the first satellite placed in space to the most recent unmanned mission to Mars.
10 Do some research and write short notes on the following:
   a Galileo Galilei
   b Isaac Newton
   c The space shuttle
   d The International Space Station.
11 a Make up five questions that you can ask your partner about the Moon.
   b Draw a diagram showing the different phases of the Moon.
   c Explain the Moon's influence on tides.
12 Write a short magazine article on the human interest in the planet Mars.

50 Topic 4: Light

What is light?

Keywords

beam, light ray, luminous, non-luminous

Did you know?

Some organisms can make and give off their own light by a chemical reaction. The making of light by a living organism is called *bioluminescence*, meaning *living light*. Fireflies and glow-worms do this to attract mates, some fish do it to lure their prey, and squid do it to scare away predators.

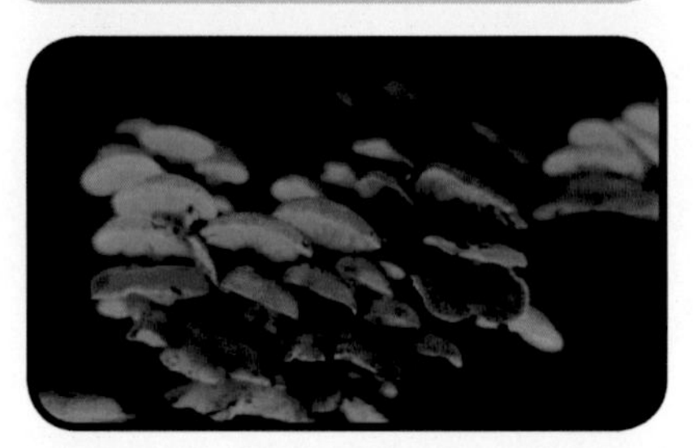

Figure 50.2 *Living lights: glow-in-the-dark mushrooms.*

Figure 50.3 *Some sources of light: lightning (above) and torchlight (below).*

Light and sight – where light comes from, and how we see

Figure 50.1 *Light enables us to see.*

The photograph above shows three candles shining in a darkened room. The light that the candles give off makes it possible for us to see. Their light enters our eyes directly so that we can see it (i.e. their light). And their light is also reflected into our eyes by the objects in the room, so that we can see the objects (e.g. the flowers and the ornament). Objects such as candles, which give off their own light, are **luminous**, while objects such as the flowers and the ornament, which do not give off light, are **non-luminous**.

The Sun is our most important example of a luminous object or body. The Sun and the stars are sources of natural light, whereas the candle and an electric light bulb are examples of artificial light because they are made by people.

Light travels fast

Light has no mass, and can therefore travel very fast. In the vacuum of space, where there is no air, the speed of light is 3×10^8 m/s. This means that light can travel seven times around Earth in one second. Light from the Sun takes about nine minutes to reach Earth.

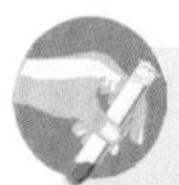

ACTIVITY

1. Classify these objects as luminous or non-luminous: candle, the Moon, star, the Sun, torch bulb, mirror, stained glass window.
2. Apart from the Sun, what other natural sources of light can you think of?

Light travels in straight lines

If you have seen a sunbeam (made by sunlight shining on dust particles in the air), or shone a torch in the dark on the smoke of a fire, you will have seen the straight path of a **beam** of light.

Later on in this topic, you will see ray diagrams showing single straight lines of light. The arrows show the direction that the light is moving in, and we call these lines **light rays**. You can think of them as very narrow beams of light.

Figure 50.4 *Tell-tale lines: sunbeams streaming through the clouds.*

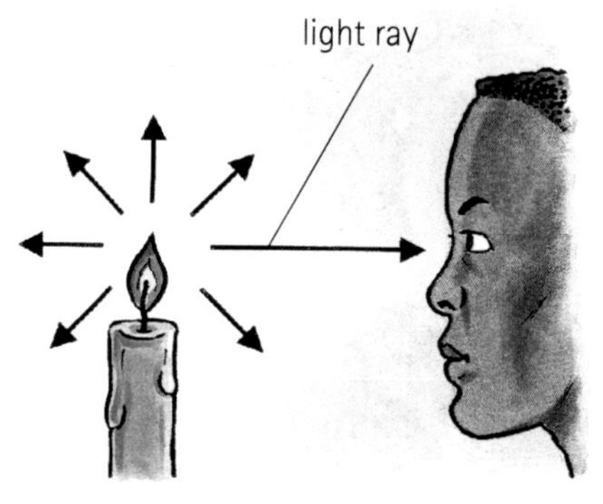

Figure 50.5 *What we mean by a light ray.*

INVESTIGATION To show that light travels in a straight line

1 Cut and fold some card stands as shown in the diagram.
2 Light a candle and place it at one end of a long piece of paper.
3 Place one of your card stands on the paper and mark the position of the slit in the card on the paper.
4 Place a second card on the paper, so that you can see the candle flame through both cards. Mark the position of this slit.
5 Repeat for three more cards.
6 Blow out the candle and remove the cards.
7 Join up the points that you have marked on the paper with a ruler. What does this tell you?

Figure 50.6 *Investigating the path of light.*

The reason you can't see what's around the corner is that light travels in straight lines. As light travels in straight lines, it follows some rules of geometry, which you will learn about later on.

Assessment

1 Explain how light makes it possible for us to see.
2 Describe how our world would be different if light didn't travel in straight lines only.

51

Light is a form of energy

Keywords

electromagnetic spectrum, wavelength

Did you know?

Light behaves in a very complicated way – sometimes as waves and sometimes as a stream of tiny particles. Particles of light are called photons.

The electromagnetic spectrum

As you have already learnt in Topics 1 and 2, light is a form of energy. The Sun warms our planet, and plants use their light to drive the chemical reaction of photosynthesis.

As a form of energy, light travels in waves, and belongs to a family of waves known as the **electromagnetic spectrum**, as shown in Figure 51.1. X-rays and microwaves are other examples of electromagnetic waves. Different wavelengths have different energies. Light waves have short wavelengths. A **wavelength** is the distance from one wave crest or peak to the next. About 1 500 of the longest light waves can fit into a space the length of 1 mm.

Ultraviolet light is a different part of the electromagnetic spectrum to visible light. Ultraviolet rays also come from the Sun, but they have shorter wavelengths than visible light. Although we cannot see them, they can damage our eyes and skin.

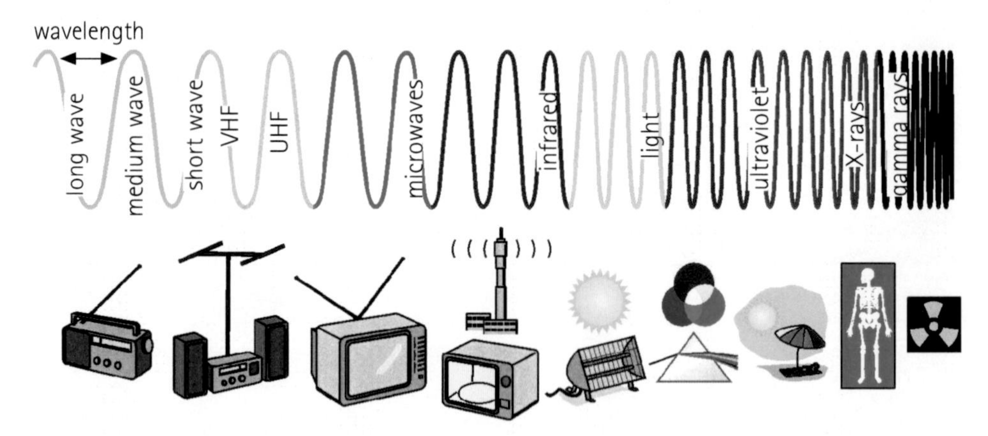

Figure 51.1 *The electromagnetic spectrum.*

Science Extra

Apart from light, all electromagnetic waves are invisible to humans. Insects such as bees can see ultraviolet light. Pit vipers have a sense organ for infrared radiation; they can find their prey in the dark by the heat that the prey's body gives off.

ACTIVITIES

1 Copy and complete this text:
Sunlight is a type of energy called … … . Microwaves and … are two other examples of this type of energy. Scientists call the light from the Sun that we can see … light. … rays are also present in sunlight; we cannot see them but they can be harmful if we are exposed to them for too long.

2 Draw a diagram to explain the term *wavelength*.

Light energy for photosynthesis

You can think of the leaves of plants as solar panels, converting the energy of sunlight into chemical energy (starch).

You have probably tested for starch in a leaf before. Now, in this investigation, you are testing whether plants can make starch without light, i.e. whether light is essential for photosynthesis.

INVESTIGATION What is the role of light in photosynthesis?

1. To block out the light, cover part of a leaf of a plant (e.g. hibiscus or geranium) with a strip of foil and sticky tape.
2. After 3–7 days, pick the leaf and remove the foil.
3. Immediately boil the leaf in hot water for 5 minutes to break down the cells.
4. Soak the leaf in heated alcohol, then in boiling water, and then stain it with iodine, as shown in the diagrams.

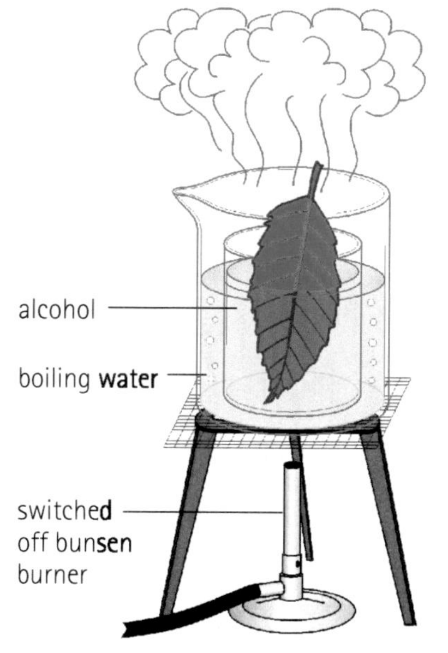

1 Soak the leaf in hot alcohol for 15 minutes to remove the chlorophyll.

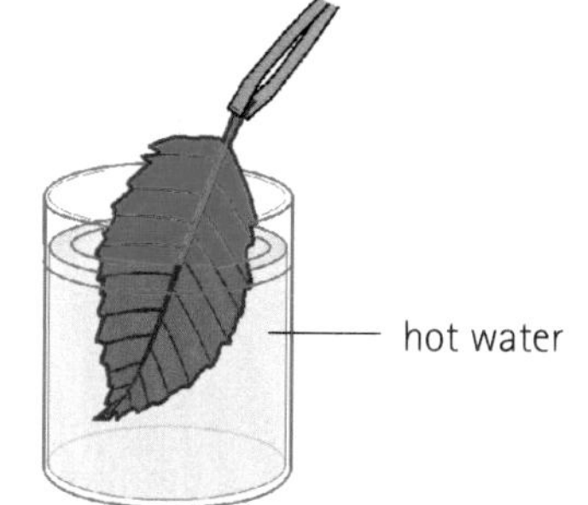

2 Rinse and soften the leaf in hot water for 10 minutes.

3 Test for starch by staining with iodine.

5. Do you see a difference in staining between the part of the leaf that was covered and the rest of the leaf? (Remember that iodine stains starch a blue-black colour.)
6. What do the results of your experiment tell you?

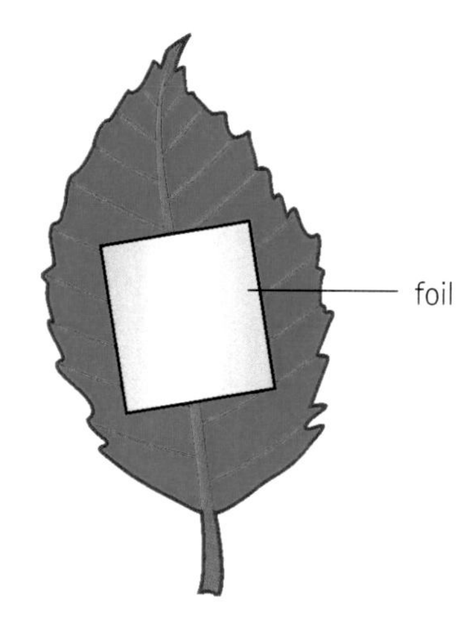

Figure 51.2 *Leaf partially covered with foil.*

Safety: Do not use alcohol near a flame or the alcohol could catch alight.

Assessment

Read 'The science of photography – drawing with light' on page 125. In what way is your iodine-stained leaf (from the experiment above) like a photograph?

52

Light and materials – transparent, translucent, opaque

Keywords

absorb, opaque, reflect, translucent, transparent

Did you know?

You cannot get a suntan through most types of glass. The ultraviolet (UV) light that causes a suntan is absorbed by the glass.

How much light do windows let through?

Glass is used for windows, greenhouses and aquariums because of its special see-through property. It is see-through because it lets the light through.

The amount of light that passes through an object depends on the material the object is made of.

- **Transparent** materials such as glass allow all, or almost all, the light to pass through.
- **Opaque** materials such as cardboard or metal block the path of light; no light passes through them.
- **Translucent** materials such as thin paper or coloured plastic let some light pass through, and block the rest.

Figure 52.1 *The tower of the Volkswagen Transparent Factory in Dresden, Germany is made of transparent glass.*

An easy way to test how much light different materials let through is to test how easily you can see through them.

Science Extra

The picture below shows 'one-way glass' on the front of a business building. This glass allows people to see through in only one direction. A special coating makes the glass very reflective – like a mirror from one side, but not from the other.

Figure 52.2 *The one-way glass of this building reflects the blue sky.*

ACTIVITIES

1. Which of these materials are transparent: tissue paper, water, glass, shade-cloth, Perspex?
2. Make a class bulletin board of materials that are transparent, translucent and opaque.
3. Conduct a survey in your kitchen or local supermarket to identify coloured glass containers and the products they contain. Suggest why these products are stored in coloured glass.

Sunlight is not only a source of light, but also of heat or warmth. In this investigation, you can compare two opaque materials to see how much light they **reflect** or **absorb**, by measuring temperature.

INVESTIGATION **How much light is absorbed or reflected?**

1 Fill two test tubes or identical glass jars with water and put a thermometer in each.
2 Cover one test tube with foil (shiny side outside) and one with black paper.
3 Record the temperature on the thermometers. They should read the same at the start.
4 Leave the two test tubes in the sun for 30 minutes.
5 Measure the temperatures again. Which test tube is hotter? Can you explain why?
6 Explain why petrol tankers and trucks that carry milk are often shiny and silver.

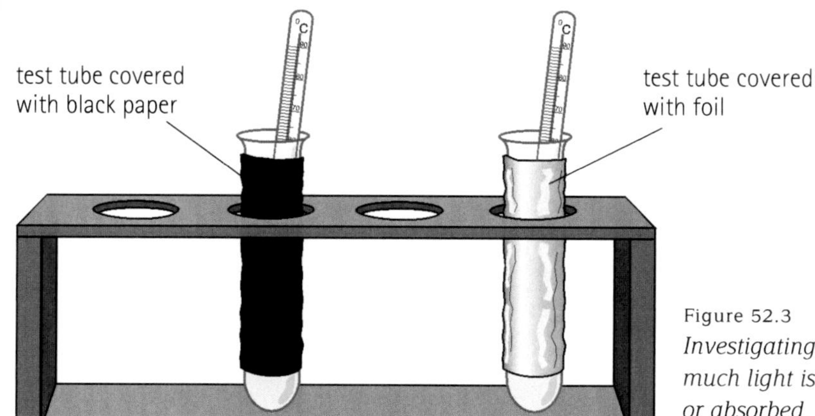

Figure 52.3 *Investigating how much light is reflected or absorbed.*

Have you noticed how sunlight can fade the colour of paint or fabric? This is because sunlight, particularly the ultraviolet part of sunlight, can break the chemical bonds of colour pigments. Exposure to sunlight for a long time can spoil food and drink products in a similar way. For this reason, some food, drink and medicine are stored in dark-coloured bottles (usually brown and green) to protect them from strong light.

Figure 52.4 *Green and brown bottles keep out some light.*

Stained glass windows

Mike Watson was a Trinidadian glass artist. His stained glass windows can be seen in churches throughout Trinidad and Tobago. Stained glass windows are often used as decoration in churches. They let the light in, but it is difficult to see through them.

Assessment

Transparent, translucent and opaque materials can be effectively used to control the light in a home. Suggest some ways these different materials can be used.

Figure 52.5 *A stained glass window.*

53

Shadows – where light can't reach

Keywords

penumbra, shadow, umbra

Did you know?

Umbra comes from the Latin word meaning *shadow* or *shade*. Umbrella comes from the Italian word *ombra* and means *a little shade*.

Shadows are made by blocking light

When you stand outside in the sun, your shadow falls in a direction opposite to where the light is coming from. Your **shadow** marks the place where light can't reach because it travels in straight lines, and it's been blocked by your body (an opaque object).

You can judge the shapes and distances of objects by the shadows they make.

Figure 53.1 *Shadows mark the place where light can't reach.*

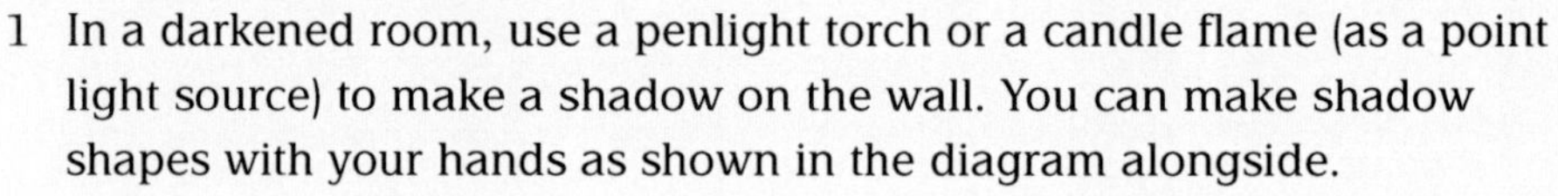

INVESTIGATION What is the difference between shadows made with small and large light sources?

1. In a darkened room, use a penlight torch or a candle flame (as a point light source) to make a shadow on the wall. You can make shadow shapes with your hands as shown in the diagram alongside.
2. Experiment with the distance of the light from the wall to get a clear, sharp shadow.
3. What do you have to do to make the shadow bigger? ... smaller?
4. Now compare the shadow you make using a lamp instead of the candle or torch.
5. Describe the difference between the shadows made with the small focused light source (the candle/torch) and the large, spread-out light source (the lamp).

Figure 53.2 *Make a hand shadow.*

Full shadow and part shadow – umbra and penumbra

You have seen in the investigation that the way a shadow forms depends on the distance between the object and the light source. The further away the object is from the light source, the smaller and sharper the shadow.

Also, when you use a small, focused light source, the shadow is very clear, with a sharp edge. When you use a large light source, the shadow has a fuzzy edge, or is surrounded by a lighter edge of shadow. The wider the light source, the more blurred the shadow.

We call the dark shadow in the centre, where no light reaches, the full shadow or **umbra**. We call the lighter region of shadow around it, where only some of the light has been blocked, the part shadow or **penumbra**.

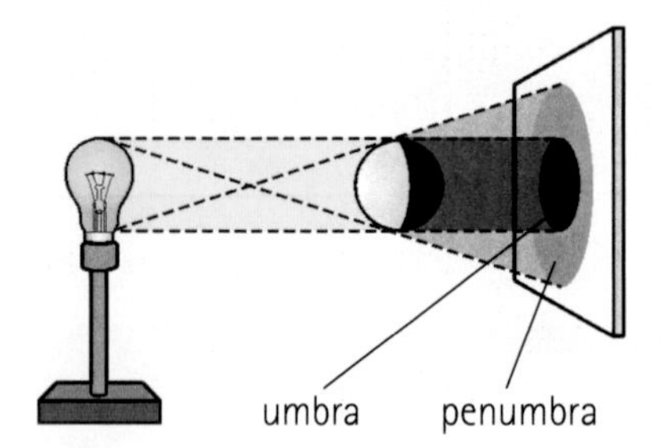

Figure 53.3 *Full shadow and part shadow.*

A solar eclipse – the Moon's shadow

You learnt about solar eclipses in Unit 45. A solar eclipse is when the Moon comes between the Sun and Earth, and forms a shadow on the Earth. The diagram shows the Moon's umbra in the centre. It covers only a small area of Earth's surface, which explains why only a small part of Earth experiences a total eclipse. The larger area is the penumbra, from where a partial eclipse is seen.

The safest way to view a solar eclipse is not to look at it directly. The diagrams in Figure 53.6 show you how to create a pinhole view of the Sun by projecting it onto a piece of paper. You can then see the shadow of the Moon across the Sun on the paper.

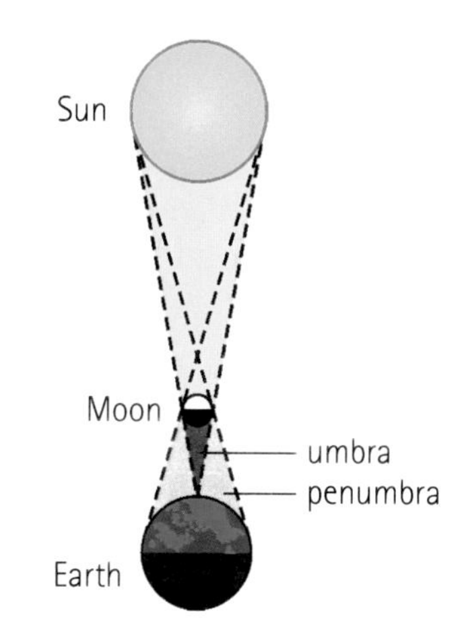

Figure 53.4 *A diagram of a solar eclipse (not drawn to scale).*

Shadows as clocks

If you face east (the direction where the Sun rises), your shadow will fall behind you in the morning, and in front of you in the afternoon. You will also see that your shadow is long in the early morning and late afternoon and shortest at midday. In this way, you can use sunlight shadows to tell the time.

ACTIVITIES

Make a simple shadow clock as follows.

1 Put a piece of paper (preferably bigger than A4) on a sunny piece of ground outside. Weight the corners of the paper down with stones. Make sure you choose a place where the Sun shines all day.
2 Push the pointed end of a pencil through the middle of the paper and into the ground.
3 Mark the line of the pencil's shadow and the time of day on the paper.
4 Predict whether the shadow will move clockwise or anti-clockwise.
5 Go back and mark the shadow an hour later. Was your prediction about the direction of the shadow's movement correct?
6 Mark the shadow at hourly intervals or whenever you can. Trace the length of the shadow and also record the time.
7 Leave the paper in its place and check the accuracy of your clock the next day. Explain why your clock would not be as accurate a month later.
8 You can also use your shadow clock as a compass. Mark on the paper where north and south lie. If you lived in the southern hemisphere, in what way would your shadow clock be different?

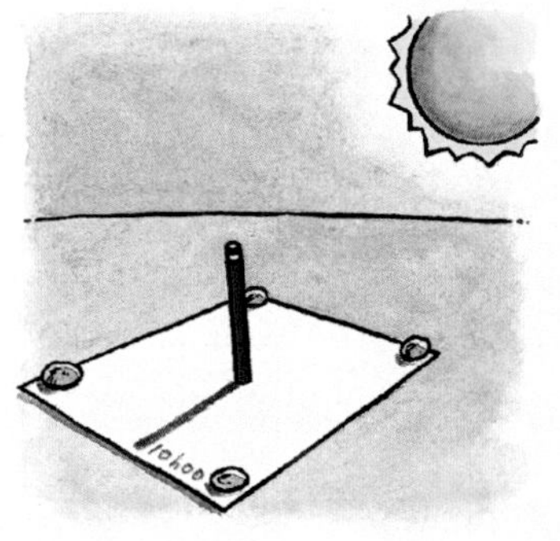

Figure 53.5 *A simple shadow clock.*

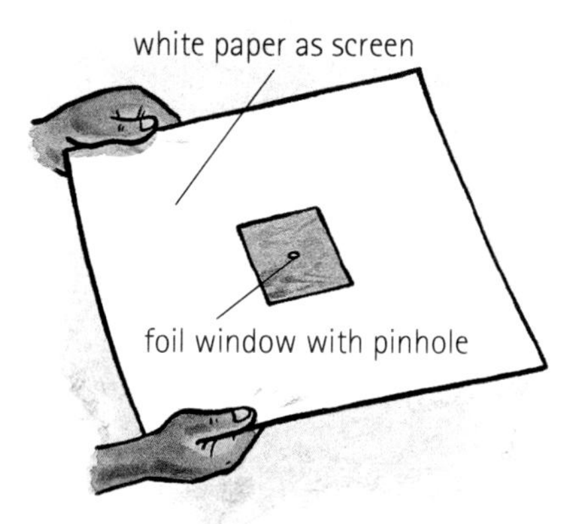

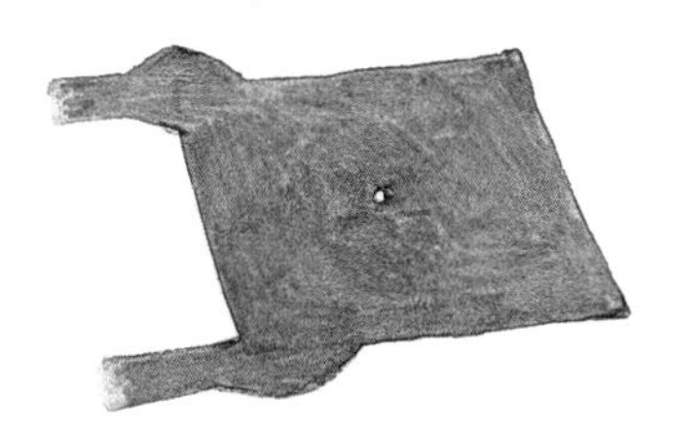

Figure 53.6 *How to view a solar eclipse without looking straight at the Sun.*

Assessment

X-rays are part of the electromagnetic spectrum. From your knowledge of shadows and further reading, explain how an X-ray of the bones in your body is formed.

54

Mirrors and reflection – when light bounces back

Keywords

angle of incidence, angle of reflection, image, kaleidoscope, mirror, periscope, reflection, virtual image

Did you know?

You can make your own simple mirror in the following way: Take a smooth piece of aluminium foil and stick it up behind a pane of glass on a window.

Reflected light

When light strikes a shiny surface, such as a mirror, the light rays bounce back in the direction that they came from. This is called **reflection**. The angle at which the light strikes the surface (the **angle of incidence**) is the same as the angle at which the light bounces off the surface (the **angle of reflection**).

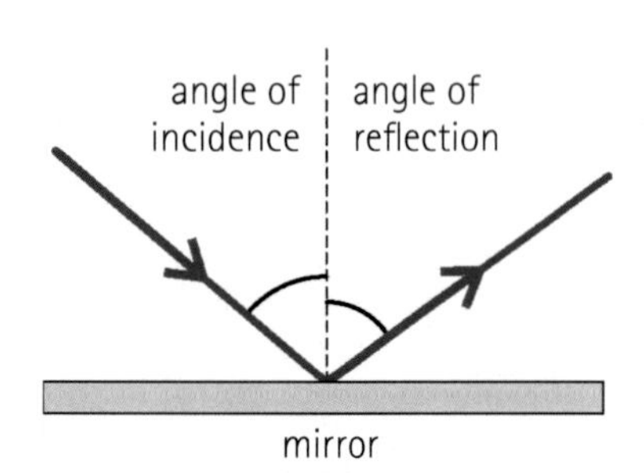

Figure 54.1 *The angle of incidence equals the angle of reflection.*

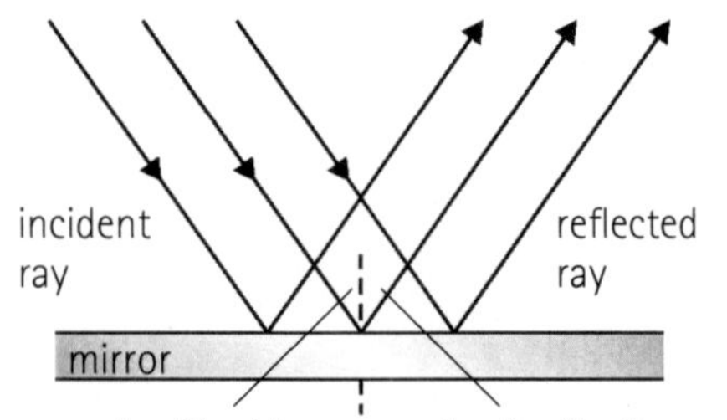

Figure 54.2 *Light rays are reflected from a mirror in parallel lines.*

Why do mirrors reflect?

You have already learnt that all opaque objects reflect light. The amount of light that is reflected depends on the smoothness and colour of the surface. **Mirrors** are special because they reflect light in such a way that we can see clear images. A white wall also reflects most of the light that hits it, but you cannot see an **image** in it. This is because if you looked at the surface of the wall under a microscope, you would see that it is very bumpy, compared to the surface of the mirror, which is very smooth. The light rays that strike the surface of the mirror are reflected in parallel lines, while those striking the wall are reflected in many different directions.

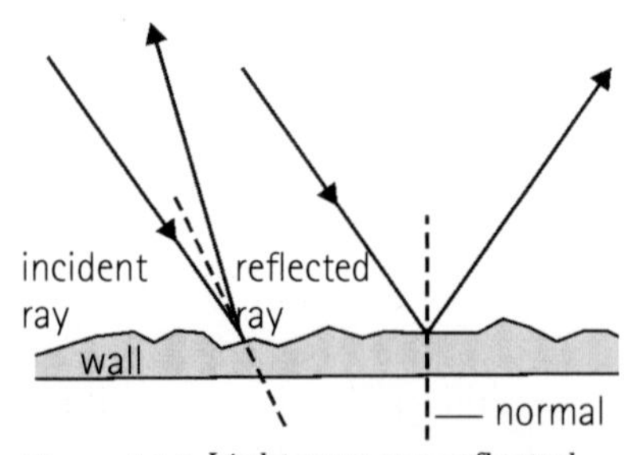

Figure 54.3 *Light rays are reflected from a wall at many different angles.*

What is special about your reflection in a mirror?

Part A

Pressing hard with a pen, write the word 'ambulance' on a piece of paper. Hold up your writing in front of a mirror. Can you read what you've written? Turn the paper the other way round so that the writing is facing you, and look in the mirror. Can you read your word?

Part B

Do this in pairs, taking turns.

Stand facing your partner, pretending that they are your reflection in a mirror. Your partner must copy or 'mirror' the movements that you make. Bend your head to the left. Which way is your partner bending their head to match your reflection? Press your right palm against your partner's palm. Which hand is your 'mirror image' using? What do you conclude about your image in a mirror?

Figure 54.4 *What is special about your reflection in a mirror?*

Mirror images

Images that you see in the mirror are back-to-front or reversed from left to right. This means that the image you see of yourself in the mirror is different from how others see you.

The image in a mirror is called a **virtual image** because, although the rays of light seem to come from it, they don't pass through or behind the mirror.

Figure 54.5 *How this person sees herself in a mirror.*

Figure 54.6 *How others see her.*

Clever ways of using mirrors

- Rear-view mirrors are used in cars so that drivers can see behind them, and you have seen in the investigation opposite why the words 'Ambulance' or 'Police' are sometimes written in back-to-front letters on the fronts of these vehicles.
- With the help of reflections from two angled mirrors in a **periscope**, you can see around corners, and submarine sailors can see above the surface of the water.
- A **kaleidoscope** makes use of three or more mirrors to reflect and multiply coloured shapes to make pretty patterns.

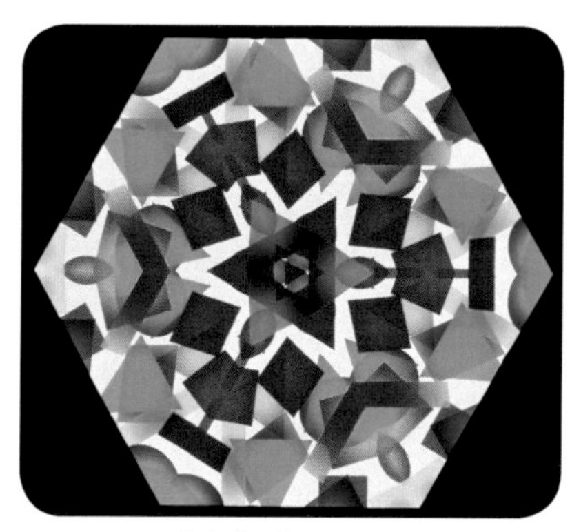

Figure 54.7 *A kaleidoscope image.*

ACTIVITY

Make your own kaleidoscope, using the diagrams below to help you. You can also research your design from a book or on the Internet.

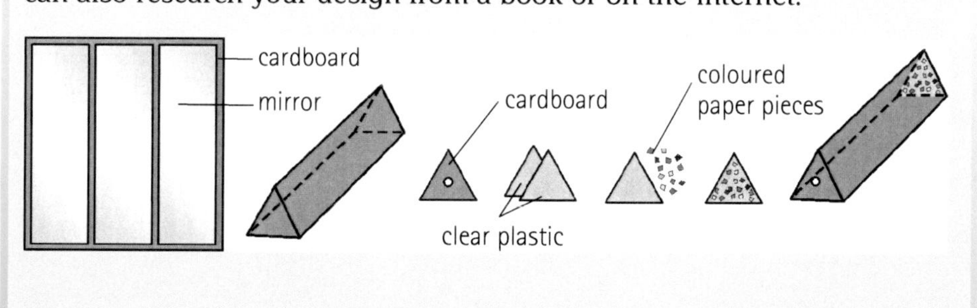

Assessment

Explain, using a diagram, why you can see reflections in still, calm water, but not in rough, choppy water.

Refraction – when light bends

Keywords

apparent depth, mirage, normal, optical illusion, refraction

Did you know?

Water refracts the light, and therefore birds such as kingfishers have to learn to adjust their aim when diving for a fish. If they aimed for exactly where in the water the fish seems to be, they would always miss.

Do I aim at the fish, in front of the fish, or behind the fish?

Light can play tricks

You've probably seen a **mirage** before – what looks like shimmering water on a road on a very hot day. It's an **optical illusion** caused by the light bending as it passes through the layer of hot air just above the tarmac. The bending of light can trick your eyes in other ways too.

Figure 55.1 *A heat mirage on a highway during summer.*

ACTIVITY

Place a straw in a glass beaker or plastic container half-filled with water as shown in the photograph. From a distance of about 30 cm, look onto the straw at the point where it is half-in, half-out of the water. Can you see how the straw looks bent? Does it look closer or further than it really is, i.e. is it bent towards you, or away from you?

Figure 55.2 *Is the straw really bent?*

Why light bends

What you are seeing in the activity above is **refraction** – the bending of light when it moves from one medium to another – in this case, from air to water. As the light rays pass from air to water, their speed is slowed down, and they bend as this happens. Think of a speeding car hitting a bank of sand at an angle. One front wheel strikes the sand before the other and therefore slows down faster, changing (bending) the direction of the car.

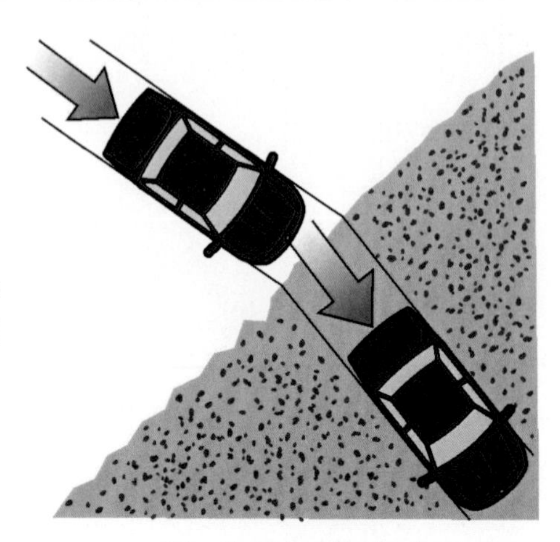

Figure 55.3 *Why light bends: think of a speeding car hitting a sandbank at an angle.*

Refraction through a glass block

The diagram on the right shows how light is refracted. As light enters the block of glass at an angle, it slows down and bends towards the **normal** (an imaginary line that is perpendicular or at right angles to its surface). When it exits the glass, it speeds up again and bends away from the normal. Although the direction of light changes when it is refracted, it keeps moving in a straight line.

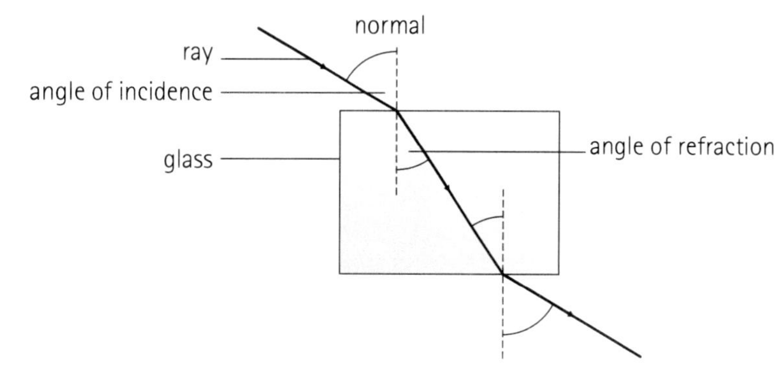

Figure 55.4 *The direction of the light changes as it goes into and out of the glass block.*

ACTIVITY

If you have a glass or Perspex block, set up an experiment in a darkened room with the help of your teacher to see what is described above.

- Pass a narrow beam of light through the block that has been placed on a piece of paper. Trace the outline of the block and the path of the light as it enters and leaves the block.

If you don't have a glass block, do the experiment as described below. You will see the effect indirectly.

- Find a glass paperweight or a glass vase with a thick base and place it on a thick solid line or clear border on the cover of a book. See how the part of the line you view through the glass is bent. You can even try this with an ice cube that's clear enough to see through.

Science Extra

Light travels at different speeds in different transparent or translucent materials. The more light slows down as it moves from one material to another, the more it bends.

Apparent depth – water can be deeper than it looks

You saw how the pencil looked bent in water. Because the light bends in the water, it seems to come from a different place, i.e. closer than it really is. In the same way, the stones on the bottom of a clear pool of water can look closer than they actually are, and this makes the pool seem shallower than it is. This optical illusion is called **apparent depth**.

Refraction is an important property of light. You will see later how lenses work by refraction.

Figure 55.5 *Objects under water look closer than they really are.*

Assessment

Without copying from the book, draw a diagram to show what happens to light when it passes through a Perspex block.

56

More refraction, and fibre optics

Keywords
endoscope, fibre optics, optical fibres, total internal reflection

Reflected instead of refracted – total internal reflection

You saw how light is refracted when it passes through a glass block. When light passes through a half sphere of glass at a sharp angle, it is also refracted. But if the angle at which the light strikes the surface is too great, all the light is reflected back though the glass, and the glass acts like a mirror. We call this **total internal reflection**.

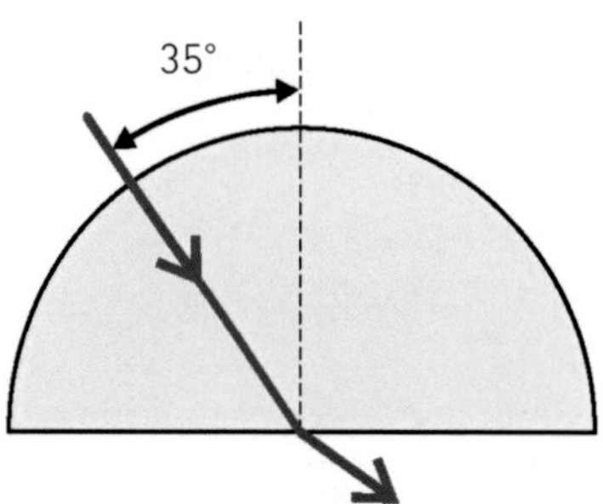

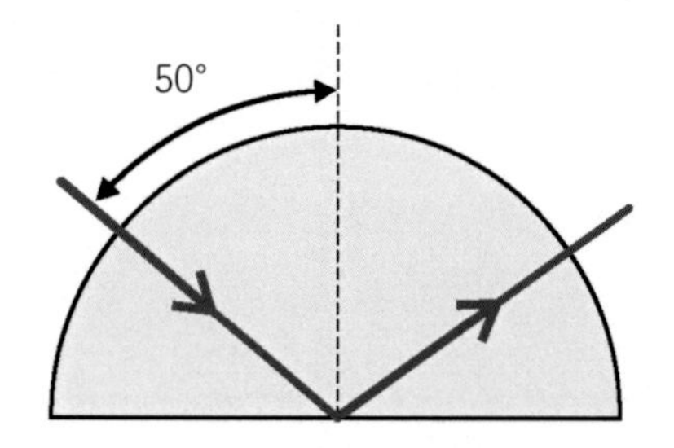

Figure 56.1 *Total internal reflection.*

Optical fibres can carry light

Optical fibres are very thin fibres of glass (as thin as a human hair), which carry light by total internal reflection. They are covered with a protective coating of plastic or resin, and because these fibres are so thin, they are flexible. They are bundled together to form fibre optic cables. **Fibre optics** is the branch of technology and engineering that makes use of optical fibres.

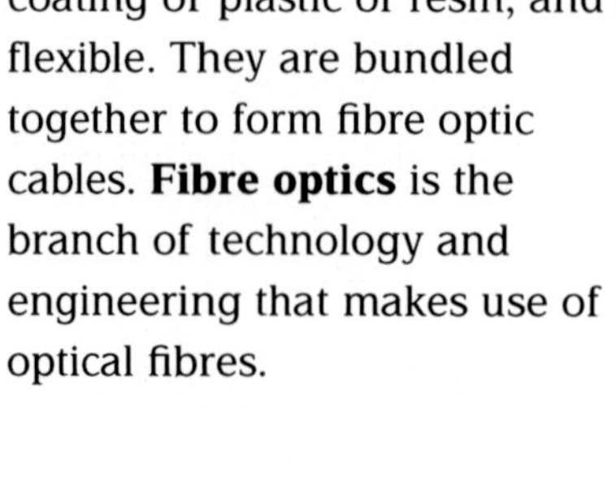

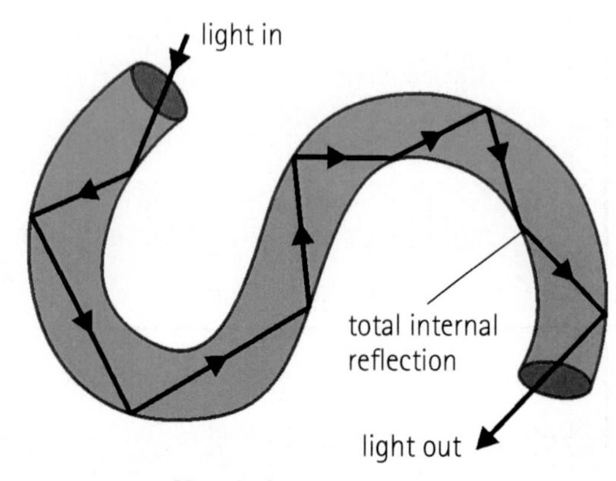

Figure 56.3 *How light moves through an optical fibre.*

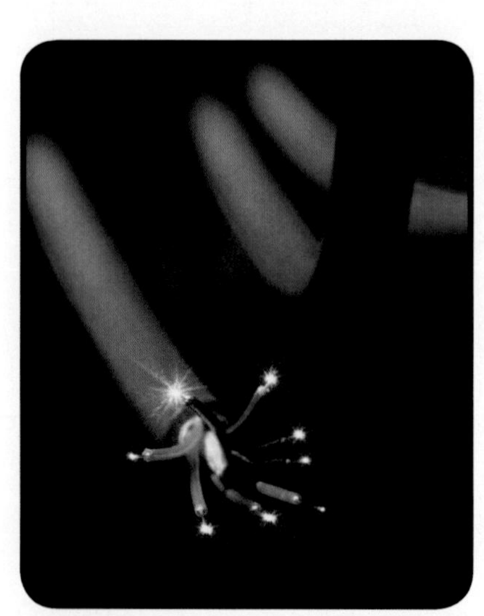

Figure 56.2 *Fibre optic cables can carry light.*

Optical fibres are used in **endoscopes** for investigative surgery, for example. Endoscopes are special light tubes that make it possible for a doctor to see into a patient's body without having to cut it open. The endoscope consists of two bundles or cables of optical fibres. Light is sent through one fibre optic cable down a patient's throat and into their stomach. The other cable is used to send images from the patient back out to a small camera. Some endoscopes even have small lasers that can carry out surgery at the same time.

Fibre optics can also carry coded signals as pulses of light, and they are therefore used in telecommunication technology to carry telephone messages and internet signals. Undersea fibre optic cables are now being used for communication networks between the continents, transmitting telephone calls and other data at high speed all over the world. Remember that light travels very fast.

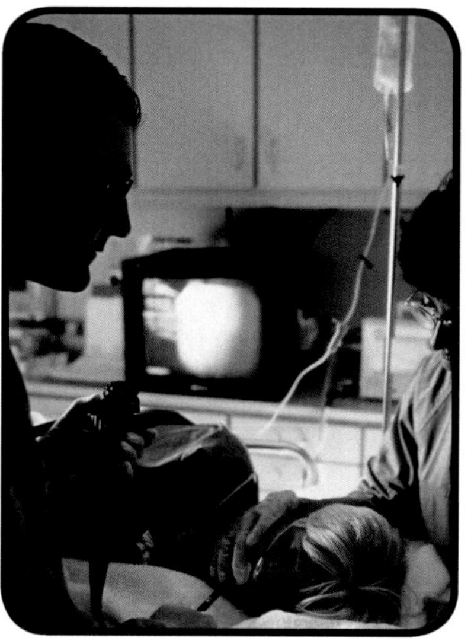

Figure 56.4 *Fibre optics are also used in medicine. This doctor is performing an endoscope examination of a patient's gastrointestinal tract.*

Figure 56.5 *A fibre optic lamp.*

ACTIVITY

In this activity you will use jelly strips to demonstrate how optical fibres carry light.

1. Make a rubbery jelly by dissolving one 80 g packet of jelly plus 60 g of gelatine in half a cup of boiling water. Pour the jelly into a rectangular container and leave it to set. (This will take about 30–60 minutes. The jelly does not need to go into the fridge.)
2. Loosen the sides of the jelly with a knife and pull it out of the container. Cut it into long strips about 15 cm long and 1 cm thick.
3. To make a narrow beam of light, cover the front of a torch with a piece of aluminium foil and make a small hole in the foil with a pin.
4. In a darkened room, hold one end of the jelly strip against the pinhole and shine the beam through the jelly strip. You should be able to see how the light passes along the length of the strip and a faint glow of light at the end of the strip where the light exits. Bend the jelly strips, and see how the light bends too.

Science Extra

Light rays can also be totally internally reflected from triangular pieces of glass pyramids called prisms. Prisms are often used in periscopes, instead of mirrors.

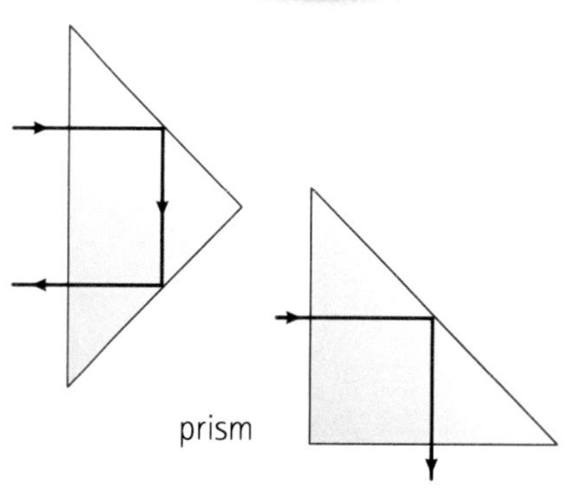

Figure 56.6 *Total internal reflection by prisms.*

Assessment

Find out more about how fibre optic technology has revolutionised telecommunications or medicine.

OR

Find out about another application of fibre optics.

57

Lenses and images

Keywords

concave, converge, convex, diverge, focal length, focal point, lens, magnifying glass, real image, virtual image

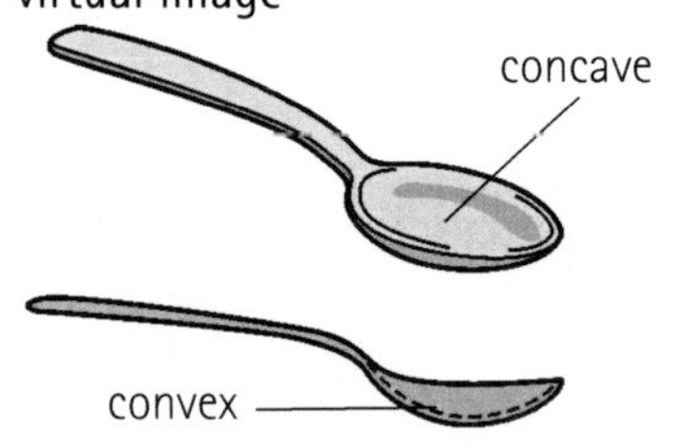

Figure 57.1 *A teaspoon has concave and convex sides.*

In Unit 55 you learnt how light is bent as it goes into and out of a medium such as glass or water, and that we call this process refraction. Now you will learn about lenses. A **lens** is a curved object that is designed to refract light and is usually made of glass or plastic.

Concave and convex lenses

Lenses are concave or convex in shape. *Concave* means *curving inwards* and *convex* means *bulging outwards*. A spoon has both a concave and a convex side. The most common type of lens is a **convex** lens, which is 'fat' in the middle and 'thin' at the edge. **Concave** lenses are the opposite shape.

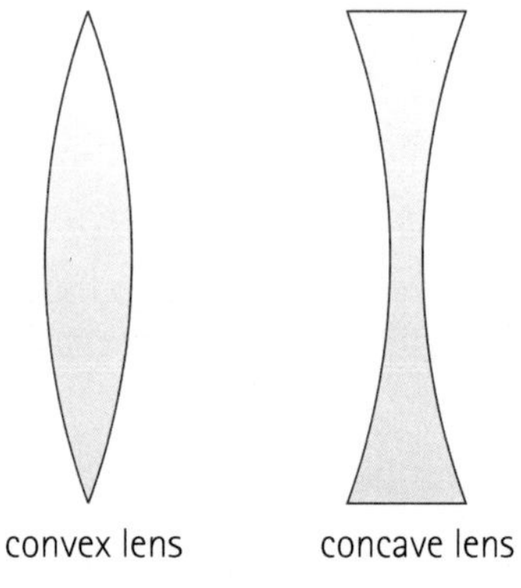

Figure 57.2 *Convex and concave lenses.*

Figure 57.3 *Making your own convex lens.*

ACTIVITY

Make your own convex lens in this way:

1. Fill a clear plastic cooldrink bottle completely full with water so that there won't be an air bubble in the bottle.
2. Hold a book with writing against the back of the bottle. What do you observe?
3. Is your lens made of plastic or water? You can test your answer by looking at the markings on a ruler through the empty plastic bottle.

Note: Keep your lens for the next activity.

In the activity above, you used a convex lens as a **magnifying glass** – a lens that makes objects look bigger than they are.

Converging and diverging lenses

When light rays pass through a convex lens, they **converge**, which means they come together in a point. Therefore a convex lens is a converging lens. When light rays pass through a concave lens, they **diverge**, which means they spread out. A concave lens is a diverging lens.

Perhaps you are wondering why, if a convex lens converges the light, it can work as a magnifying glass as you saw in the activity above. You will find out the answer soon. First, you need to learn about the focal point and focal length of a lens.

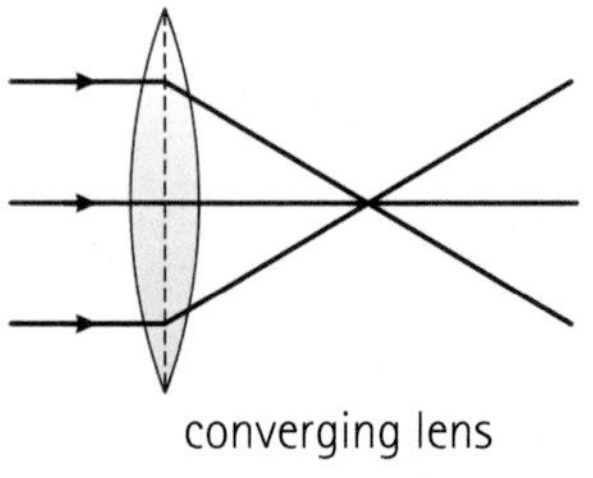

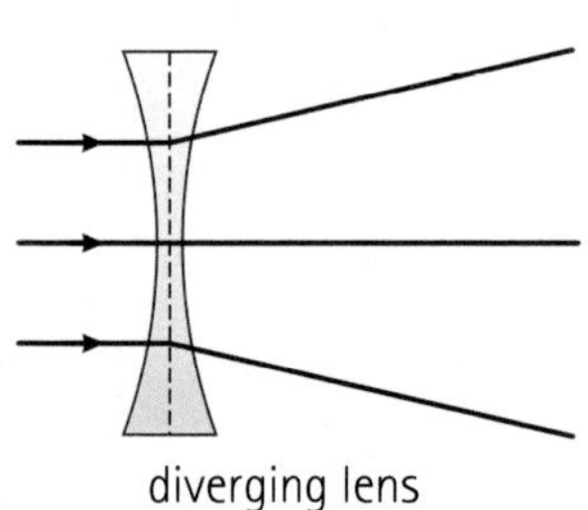

Figure 57.4 *Converging and diverging lenses.*

Focal point and focal length

A converging lens makes the light converge at a single point called the **focal point**. Beyond this point, the light rays cross over and diverge. The distance from the centre of the lens to the focal point is called the focal length. The **focal length** is a measure of how strongly the convex lens focuses the light, and it depends on how thick the lens is. The thicker the lens, the more it bends the light.

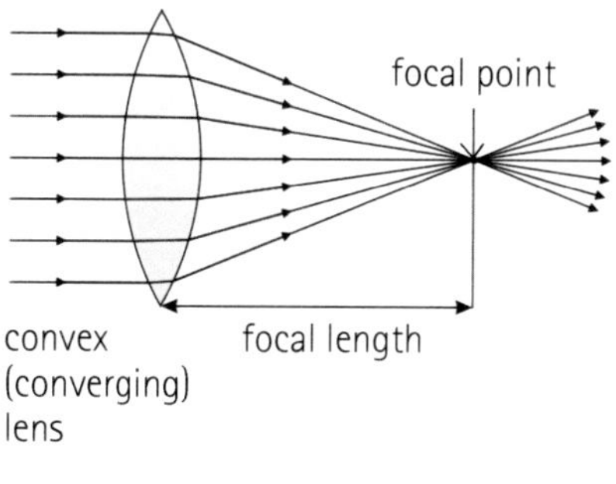

Figure 57.5 *Focal length and focal point of a convex lens.*

Viewing close and distant objects through a convex lens

The image formed by a convex lens when an object is far away is very different from the image formed when an object is close. When light rays from a distant object pass through a convex lens, they converge to form a small upside-down image. It is a **real image** because the light rays from the object pass through the lens and can be shown on a screen.

If an object is close to a convex lens, and the distance between them is less than the focal length of the lens, the rays of light can't converge to form a real image. Instead, to the viewer, the rays seem to come from a position further away, which makes the object look bigger than it is. It is a **virtual image**, because the light rays don't really come from where they seem to.

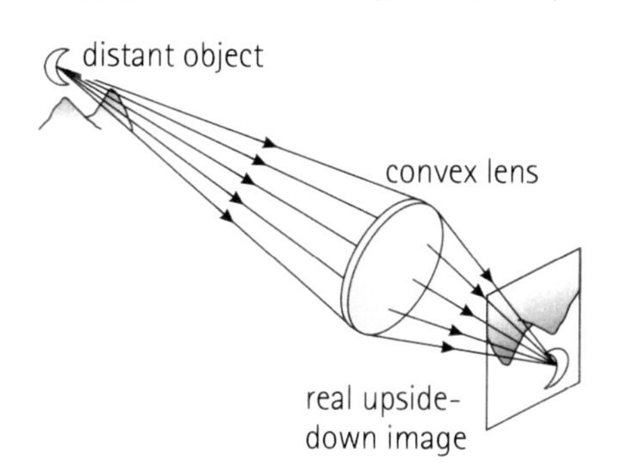

Figure 57.6 *Viewing a far-away object through a convex lens.*

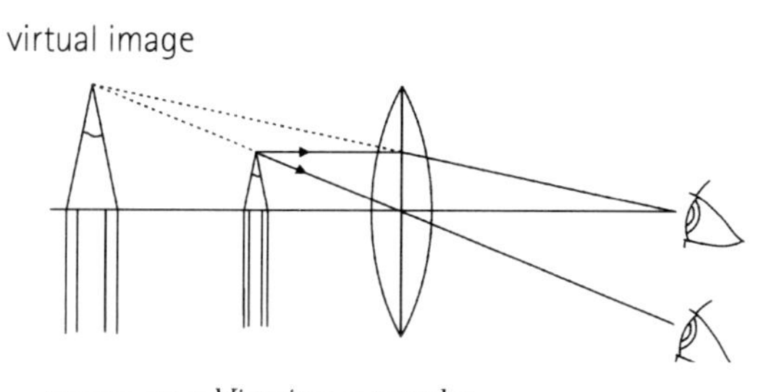

Figure 57.7 *Viewing a nearby object through a convex lens.*

ACTIVITY

Test what you have learnt about looking at distant objects through convex lenses, using your water bottle again. Stand at a window where you can see treetops or the roofs of buildings and the sky. Hold the bottle on its side near to your eyes. Can you see an upside-down image? You will see this effect more clearly in the next unit. You can also test this by looking at close and distant objects through a magnifying glass.

Assessment

1 Is a magnifying glass made of a concave or a convex lens? Is it a converging or diverging lens? Explain why it can be used to burn a hole in a piece of paper on a sunny day.

2 Complete this sentence: The thicker the convex lens, the … (shorter/longer) the focal length.

58

Lenses and telescopes

Keywords

eyepiece lens, objective lens, optical instruments, telescope

Lenses are used in **optical instruments** such as microscopes, binoculars and **telescopes**. Their purpose is to make objects that are small or far away more visible. See how a lens can focus an image in the next activity.

Figure 58.1 *Using a convex lens to focus an image.*

ACTIVITY

Work in pairs and use a convex lens to focus an image as follows:

1 Cover the front of a torch with a piece of foil. Take a pin and poke a 'face' in the foil – two holes for the eyes, and a narrow slit for the mouth.
2 Place the torch on a desk in a darkened room so that the face is the right way up and the torch is shining 0.5–1.0 m away from the wall. At this stage, you should see only a faint fuzzy beam of light on the wall.
3 Hold a magnifying glass or a convex lens in front of the torch. Slowly move the lens towards or away from the torch, until the image of the face on the wall is focused. What do you notice about the image?
4 Turn the torch so that the picture of the face is upside down, and repeat.

The focal length of a lens is a measure of how strongly it focuses light. The shorter the focal length, the more powerful the lens. As you'll see in the next activity, you can measure the focal length of a lens by focusing light from a distant object.

Figure 58.2 *Measuring the focal length of a convex lens.*

ACTIVITY

Measure the focal length of a convex lens this way:

1 Hold a lens or a magnifying glass so that the Sun shines through it. DO NOT look at the Sun through the lens – it will damage your eyes. (If the lens has a flat side, the flat side must face the Sun.)
2 Focus the light on a piece of paper to make the smallest dot of light you can.
3 Measure the distance from the lens to the paper. This is the focal length. Warning: You will need to do the activity briskly. If you focus the light on the paper for too long, the paper will catch fire.

Science Extra

Microscopes and telescopes have two or more lenses that together refract the light to the eye, and so magnify the object. A microscope has its own light source that brightens the object enough to see it. A telescope cannot do this, and must therefore have a larger lens to allow more natural light in from distant objects.

Telescopes

You saw in Topic 3 how telescopes are used to help us look at the planets and stars. A telescope gathers and focuses light from distant objects. Turn back to the previous unit to see how a convex lens forms an image of a distant object.

ACTIVITY

To make your own simple telescope, you will need: 2 cardboard mailing tubes (one fitting snugly just inside the other), 2 convex lenses (one big, one small), thick cardboard, tape and scissors.

Follow these instructions:

Work out the focal length of your lenses

A basic telescope has two lenses:

- a larger, **objective lens** that focuses the image from the distant object; it has a longer focal length
- a smaller **eyepiece lens** that magnifies the image; it has a shorter focal length.

Use the method in the second activity on page 120 to measure the focal length of each of your two lenses.

To calculate the magnification of your telescope:

$$\text{magnification} = \frac{\text{focal length of objective lens}}{\text{focal length of eyepiece lens}}$$

Make the body of your telescope

Cut the two cardboard tubes to the correct length (see the box in the margin) and carefully slide the one into the other.

To work out the correct length for the tubes:

1. Add the focal lengths of the two lenses, and divide by 2, e.g. $(25 + 5) \div 2 = 15$
2. Now add 5 cm to this length, e.g. $15 + 5 = 20$
3. Cut both the tubes to this length.

Make the lens holders

For each lens holder:

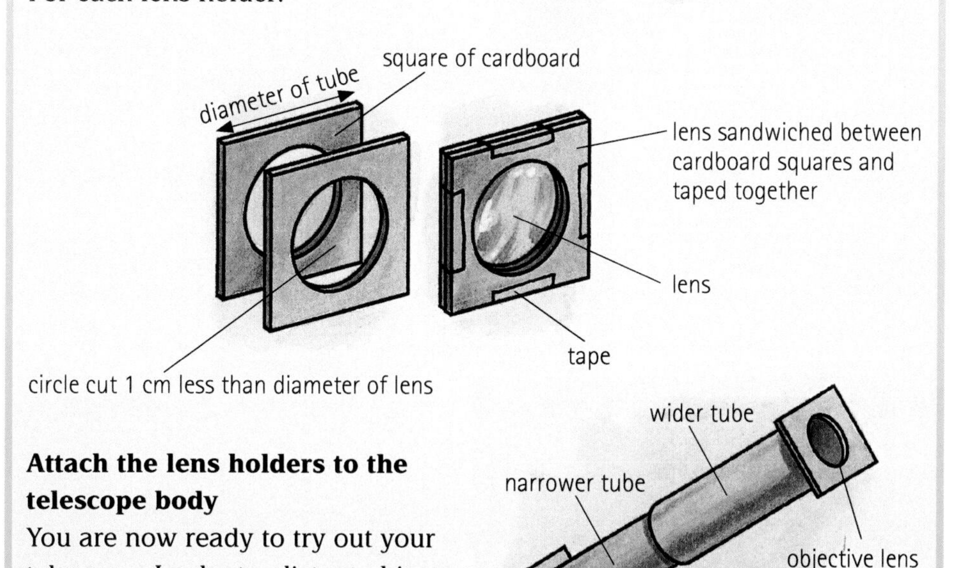

Attach the lens holders to the telescope body

You are now ready to try out your telescope. Look at a distant object. Slide the cardboard tubes until you get a clear image.

Safety: Never use your telescope to look at the Sun or other bright lights. The telescope concentrates the light from a larger area into your eye, and this could damage it.

Assessment

How well did your telescope work? Compare your telescope with those of others in the class. What could you do to improve the design?

59

How do your eyes work?

Keywords
cones, cornea, iris, lens, long-sighted(ness), pupil, retina, rods, short-sighted(ness)

Did you know?

Cheap sunglasses can damage your eyes. The darkness of the glasses makes the pupils widen, letting harmful radiation, not blocked by the sunglasses, in. Make sure you get glasses that block harmful UV-A and UV-B radiation.

The structure and function of the eye

The human eye is a tough jelly-filled ball that sits in a bony socket. It is also a complex organ with its own lens system. The diagram on the right shows the structure.

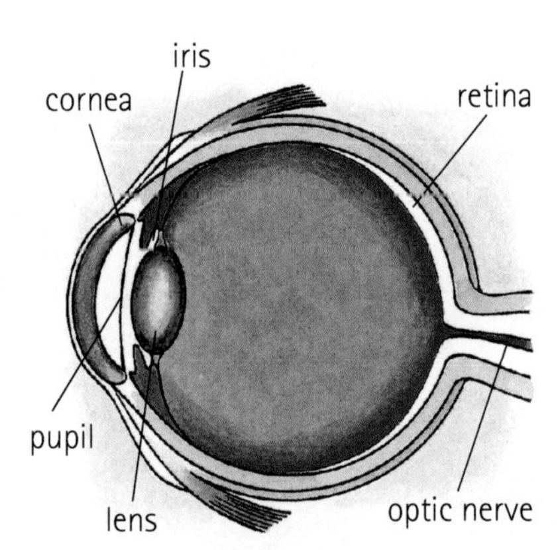

Figure 59.1 *The human eye is a complex organ.*

The iris controls how much light enters the eye

Light enters the eye through an opening called the **pupil**, and the eye can adjust to changing amounts of light. In bright light, the **iris** (the coloured ring around the pupil) gets bigger and the pupil smaller, letting less light in. When it is dark, the iris makes the pupil bigger to let in more light.

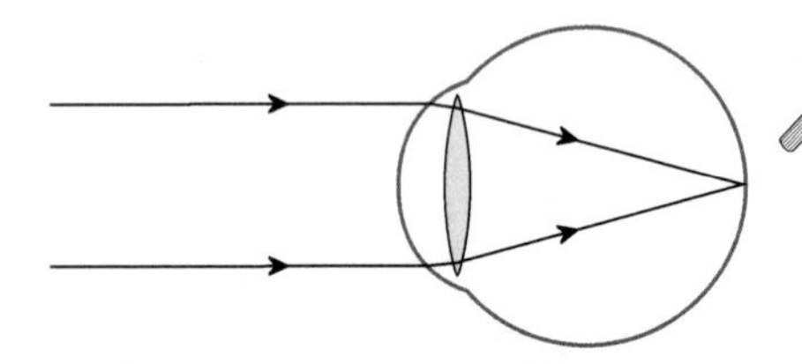

Figure 59.2 *Light from a distant object only needs to be refracted (bent) a little bit to make an image. The lens flattens.*

INVESTIGATION What happens to the pupils of the eye in bright light?

Have you ever noticed that if you come into a dark room after being in the Sun, it takes your eyes a while to adjust to the dim light? Do this investigation in pairs, taking turns. If you do it on your own, you will need a mirror.

1. After 1–2 minutes in a darkened room, look closely at the pupils of your partner's eyes.
2. Then switch on the lights or open the window blinds and see what happens to the pupils.
3. What do you observe?

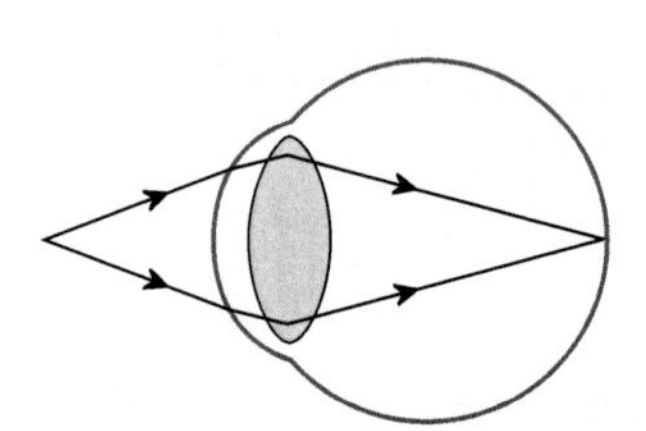

Figure 59.3 *Light from a close object needs to be bent a lot. The lens bulges to cause more refraction.*

The lens focuses the light by adjusting its shape

The **cornea** is the protective surface of the eye, and together with the **lens**, it bends the light rays that reach the eye. The convex lens focuses the light so that the image falls on the **retina** at the back of the eye. The lens is made from an elastic substance and can change its shape. Have a look at Figures 59.2 and 59.3, showing how the eye focuses on a close object and on a distant object.

An upside-down image forms on the light sensitive retina

The surface of the retina is made up of millions of light sensitive cells called cones and rods. The **rods** are sensitive to the amount of light (light intensity), and the **cones** are sensitive to colour. The upside-down image on the retina is transported by the optic nerve to the brain, which interprets the signals, and creates an image that is the right way up.

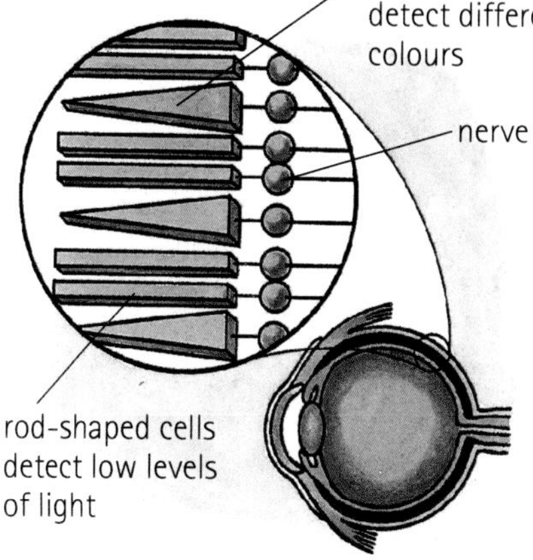

cone-shaped cells detect different colours

nerve cells

rod-shaped cells detect low levels of light

Figure 59.4 *The retina is covered in cones and rods that are sensitive to light.*

Short-sightedness (myopia) and long-sightedness (hypermetropia)

Many people have eyes that do not see perfectly. But most problems with eyesight can be easily corrected with glasses or contact lenses.

A **short-sighted** person can only see objects close up, but not far away. Their eyeballs are too long (or their lenses too 'fat'), and so light from distant objects is refracted too much, and the focused image forms in front of the retina. Glasses with concave lenses spread the rays out so that they now focus further back, and the image forms directly on the retina. Short-sighted people can also have their vision permanently corrected with laser surgery. In the surgery a small layer of cornea in front of the lens is removed.

A **long-sighted** person can see far objects clearly, but not close up ones. This is because their eyeballs are too short for the lens system, and the focused image falls beyond the retina. Glasses with convex lenses bend the rays more for better vision. As people get older, their eye lenses become harder and less flexible, and they become long-sighted. Most people need to wear glasses for reading by the time they are fifty.

Science Extra

It is difficult to see well underwater because the human eye has difficulty focusing the light. The lens is unable to bulge enough to bend the light rays in water into the eye. Sea animals such as seals are able to change the shape of their lens more than humans, so that they can see well both above and below the water.

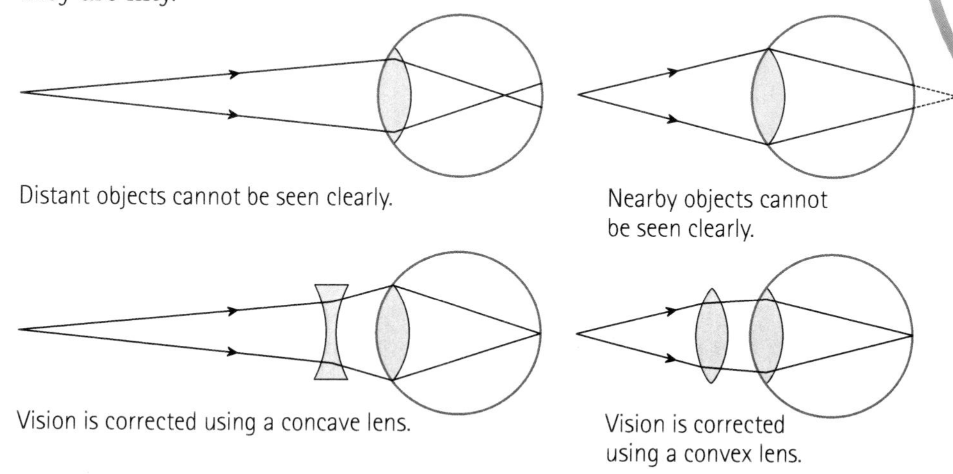

Figure 59.5 *Short-sightedness*

Figure 59.6 *Long-sightedness*

Assessment

If you can, have a look at a 3D model of the human eye. Can you identify all of the parts that you see in the diagram on page 122?
OR
Research laser surgery used to correct eye defects.

60

The camera obscura and the pinhole camera

Keywords

aperture, camera, lens

Did you know?

The first camera was designed in 1839 and was the work of the inventor Louis Daguerre of France.

Comparing the camera and the eye

A **camera** works in a similar way to the eye. How is the camera like the eye?

ACTIVITY

Now that you've learnt about how the eye works, look at the diagram of the camera. Identify which parts of the camera match the following parts of the eye, giving reasons: pupil, iris, lens, retina.

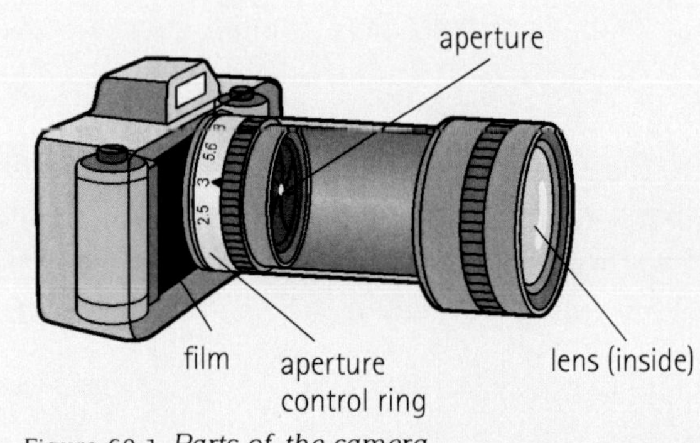

Figure 60.1 *Parts of the camera.*

Science Extra

The sensor inside a digital camera is sensitive to the different colours and intensities of the light falling on it, and sends this data to a memory card so that it can be transferred to a computer screen to be displayed. This is similar to the optic nerve carrying the light impulses to the brain.

The difference between the eye and the camera is that the camera has a rigid **lens**, which focuses by moving in and out. Some cameras have a fixed lens and therefore the focal length cannot be adjusted. These cameras can't take close-up shots – they can only photograph objects at a distance.

The camera obscura

The camera obscura, which means *dark room* or *chamber*, was the forerunner of the cameras we use today. A camera obscura is a dark box or room with a small opening or hole in one end. When light from outside enters through the hole, it forms an upside-down image of the outside view on the opposite wall.

How does this work, and why is the image upside down? The small opening in the box or room orders the jumble of light coming in. Similar to a lens, the hole makes the light rays converge and cross over. The rays then re-form as an upside-down image. The smaller the hole, the clearer the image.

The Egyptian scientist Ibn Al-Haytham (Al-Hazen) described this in his writings in the 10th century, although he did not call it a camera obscura. Modern cameras work on the same principle: light from an object goes through a small hole called the **aperture** in the front of the camera and forms an upside-down image on a light-sensitive film inside the camera.

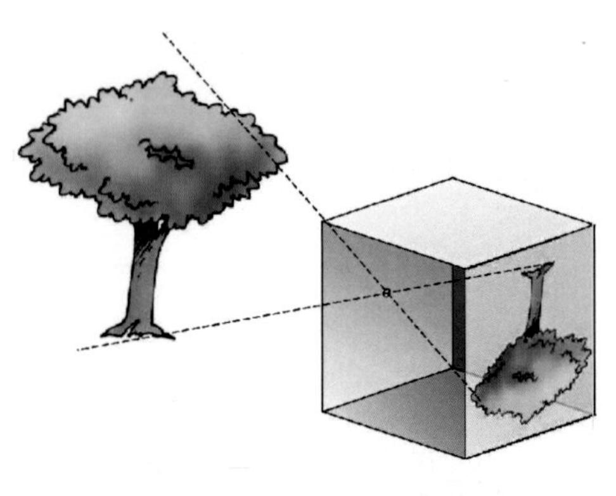

Figure 60.2 *How the pinhole camera works.*

The pinhole camera

A pinhole camera is a camera without a lens. It works in the same way as a camera obscura. It is easy to make a pinhole camera yourself. The next activity shows you how.

ACTIVITY

You will need: an empty Pringle's chips tin/tube (170 g) with a clear plastic lid, a tissue, a ruler, a pen, a bread knife, sticky tape, foil, greaseproof paper, scissors, a drawing pin and a thin knitting needle.

1 Wipe the inside of the tube clean with a tissue.
2 Measure 5 cm from the bottom of the tube, all the way around, and saw off this end with a bread knife.
3 Pierce two holes in the centre of the metal bottom of the short piece of tube – one with a drawing pin and one with a knitting needle. Cover the larger hole with a small piece of foil and sticky tape.
4 Place a square piece of greaseproof paper over the open end of the short tube (waxy side up), place the lid on top and tape in place.
5 Put the longer piece of the tube back on top of the shorter piece and tape the two pieces of tube together.
6 To make the tube light-proof, cover it with foil, shiny side out.
7 Point the pinhole of your camera to a window, and look at the view outside though the open end. Cup your hands around your eye to keep out the light. What can you see on the screen made with the lid?

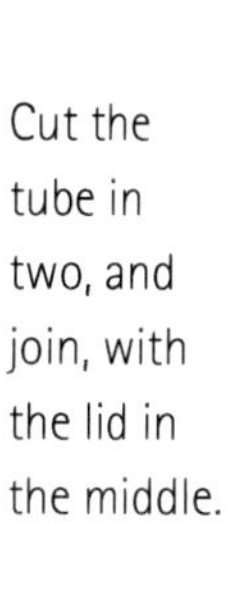

Figure 60.3 *Making a pinhole camera.*

You can now use your pinhole camera to do the following investigation.

INVESTIGATION Testing your pinhole camera

How does the size of the pinhole change the image?
Open the larger pinhole and cover the smaller one. How does this change the image?

What is the effect of using a lens?
Keeping the large hole open, hold a convex lens over the front of your camera. Move it nearer to and away from the pinhole. How does the use of a lens affect the image?

The science of photography – drawing with light

If you put light-sensitive film or paper inside your pinhole camera where the image forms, and then develop it in the dark, you would have a photograph. The word *photography* means *drawing with light*.

Camera film is coated with silver bromide – a chemical that is sensitive to light. A chemical change takes place in the parts of the film that are exposed to light, and a 'negative' image is made when the film is developed – the parts of the film where the light was the brightest become the darkest. To make a 'positive' print, light is then shone through this negative onto light-sensitive photographic paper.

Figure 60.4 *This photograph was taken with a pinhole camera.*

Assessment

Draw a diagram showing how light passes through a pinhole camera. Use this to explain why the images are upside down.

61 Splitting light into the colours of the rainbow

Keywords

colour spectrum, dispersion, prism, rainbow, Rayleigh scattering

Did you know?

Have you ever seen the green flash of the Sun as it slips below the horizon at sunset? It is caused by the same refraction and scattering effects that make the sunset red.

The many colours of sunlight

White light from the Sun is a combination of seven colours – red, orange, yellow, green, blue, indigo and violet, plus all the colours in between. You have seen all these colours in a rainbow.

When light passes through a triangular-shaped piece of glass called a **prism**, it splits light into a range of colours called the **colour spectrum**. This splitting of white light into colours is called **dispersion**.

Dispersion happens because of refraction. The different colours of light have different wavelengths, and so they are bent at different angles and travel through the glass at different speeds. Red has the longest wavelength (about 700 nm) and is bent the least, while violet light has the shortest wavelength (about 400 nm) and is bent the most. The different colours are refracted even more as they leave the prism because of the prism's triangular shape.

Science Extra

Lasers produce focused, narrow beams of light, usually of one colour or wavelength only. Low-power lasers are used in barcode scanners, laser printers and CD players. A laser beam 'reads' the music data on a CD. A lens and prism system is used to shine the laser beam onto the disc and then reflect the beam off the disc to a light-sensitive detector.

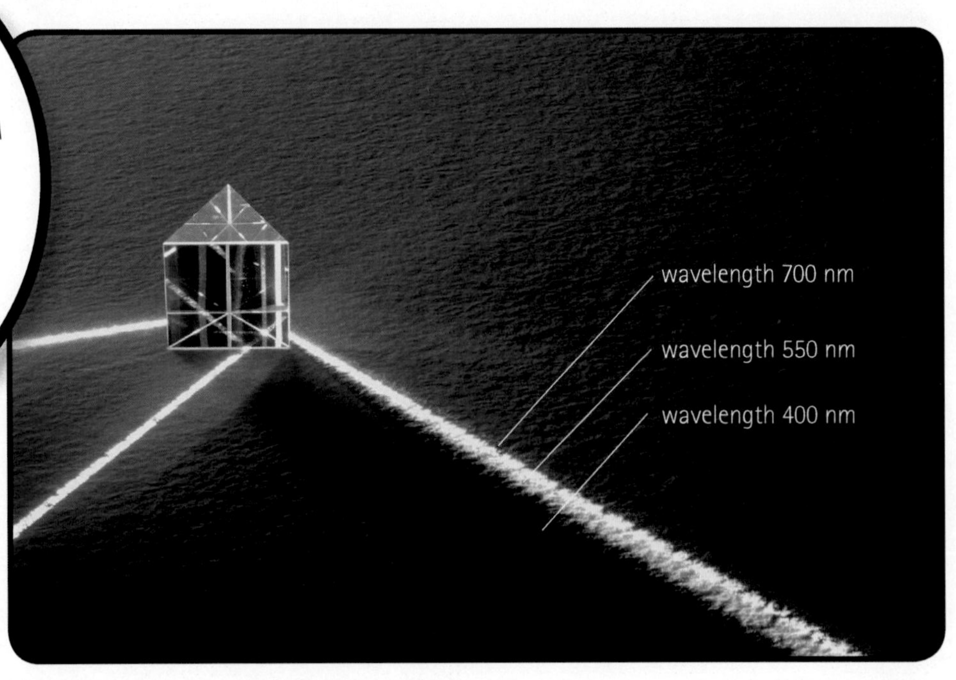

Figure 61.1 *A prism can split light into colours.*

Figure 61.2 *The laser beam reflects off the disc to a light-sensitive detector.*

If you have a prism, your teacher will show you how it separates light into colours. If you don't have a prism, you can easily see the colour spectrum reflected off a compact disc (CD) (see the next activity). The CD is made of plastic that is coated with aluminium. Light is reflected off the concentric rings or ridges in the metal of the CD.

ACTIVITIES

Make a CD colour spectrum

1. Hold a CD with its blank side (the side with no picture or writing) towards you, and tilt it in the light. Can you see each of the seven main colours of the spectrum?
2. Take the CD and a piece of white paper outside and reflect the colours off the CD onto the paper. You should be able to make a 'rainbow'.

Make a Newton disc

1. Copy the template on the right onto a piece of paper and colour it in with paints, pastels or crayons.
2. Cut out the disc and stick it onto the end of a kebab stick or a knitting needle. You will need to hold it in place with some glue.
3. Once the glue is dry, rub the stick between your hands to spin the disc as fast as you can. See how the colours blur towards white. (You can see this effect best if you can set the disc to spin on the bobbin of a sewing machine or an old film reel.)

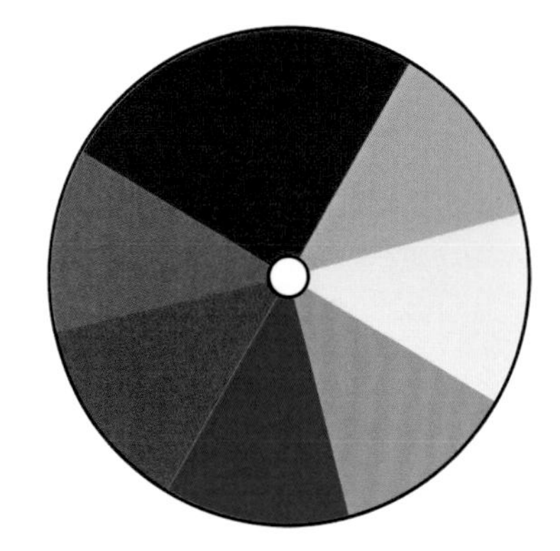

Figure 61.3 *A Newton disc.*

What makes a rainbow?

Rainbows are made by the refraction of sunlight in water droplets in the air after rain. Each water droplet acts like a tiny prism. Light that enters the droplet is refracted. It is then reflected from the back of the drop by total internal reflection, and refracted again as it leaves. The colours of light are separated in this way.

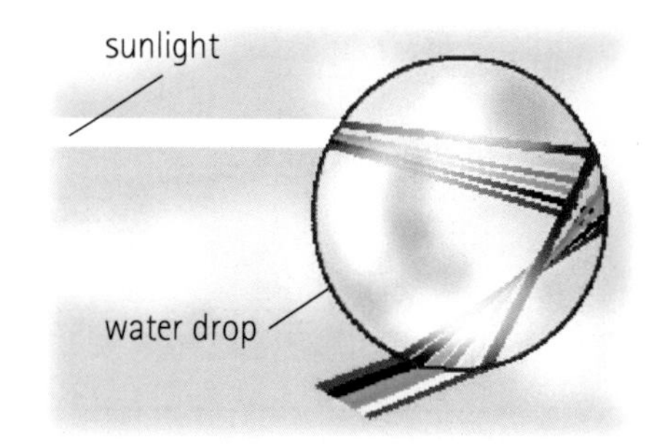

Figure 61.4 *A water drop acts like a prism to form a rainbow. Light is split into colours when it is reflected inside the drop, and is refracted more when it leaves the drop.*

Why is the sky blue and the sunset red?

Have you ever wondered why the sky is blue? The blue end of the colour spectrum (blue, indigo and violet light) has the shortest wavelengths. As the sunlight passes through the atmosphere, only the blue light is absorbed by the molecules of oxygen and nitrogen in the air, and then scattered in all directions. This scattering of blue light is called **Rayleigh scattering**.

Red light, at the other end of the colour spectrum, has the longest wavelength, and so is not easily scattered by the atmosphere. At sunrise and sunset, because of the low angle of the Sun, the light's path through the atmosphere is the longest. It is only the red light that isn't scattered; it travels uninterrupted through the air, and reaches our eyes.

Figure 61.5 *Blue light has a shorter wavelength than red light and is more easily scattered.*

Assessment

Explain to someone younger than you how a rainbow forms. You can do this by diagrams and/or demonstration (e.g. spray a fine mist of water from a hose in the Sun). See how well you can get them to understand your explanation.

62

Seeing colours

Keywords

filter, pigment, primary colour, secondary colour

Did you know?

Birds have four types of colour-sensitive cone cells. Most mammals, including dogs and cats, cannot see colour as well as we do. Dogs and cats can't see red.

Mixing the primary colours

Blue sky, green leaves, red flowers. Blue, green and red are the **primary colours** of light. If we mix two primary colours together, we get a **secondary colour**. If you have ever done stage lighting for a school play, you will already know this:

red + green = yellow
red + blue = magenta
green + blue = turquoise (cyan)

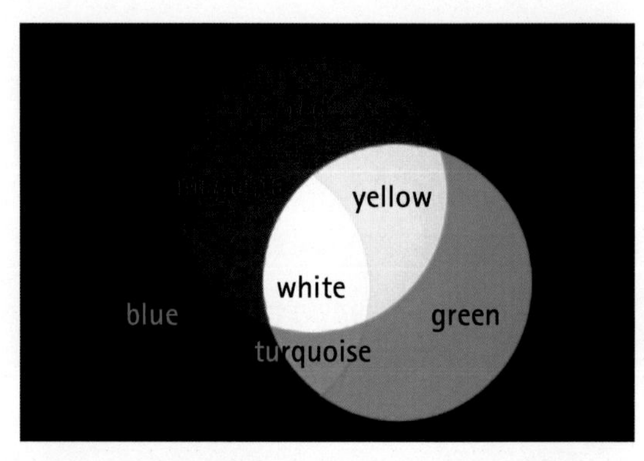

Figure 62.1 *All colours are made up of just the three primary colours: red, green and blue.*

Figure 62.1 shows the primary and secondary colours for light. (The primary and secondary colours for the mixing of paints and dyes are different; for paints, yellow is one of the primary colours.)

Figure 62.2 *Carnival colours.*

ACTIVITY

To see the secondary colours for yourself, you will need: two to three torches and red, green and blue transparent sweet wrappers or Cellophane. (You can also use crinkly plastic shopping bags with coloured logos as long as they're made of thin enough plastic.) Then in pairs or a small group:

1. Cover the front of one torch with a red sweet wrapper or Cellophane, and the front of the other torch with a green wrapper.
2. Shine these closely onto a sheet of white paper or a wall. (It works better if you do this in a room that's not too brightly lit.) You should be able to see a red spot of light and a green spot of light.
3. Now make the two colours of light overlap to see the secondary colour. (You might need to experiment with the distance that you hold the torches from the paper and the brightness of the room to see this best.)
4. Repeat this for the other colour combinations.
5. Can you predict what colour you will get when you mix red and green and blue light?

Seeing in three colours

You learnt in Unit 59 that the retina of your eye has cone cells. There are three types of cone cell, and each type is sensitive to a different colour of light – red, green or blue. This means that you see all colours in terms of red, green and blue. Your brain senses a particular colour by the way it stimulates these three types of cone cells.

People who are colour-blind have trouble telling the difference between colours – either red and green, or blue and green. An average of 7–10% of people are red-green colour-blind, and almost all of them are male. This is because the genes for the red and green cone cells are on the X chromosome. Men have only one X chromosome, but women have two. So if one of a woman's X chromosomes lacks these genes, the other X chromosome is likely to have them.

Science Extra

A television or computer screen uses the three primary colours in different combinations to produce a multicoloured picture. The screen is divided into small block or dots called pixels, and these pixels are coloured in by red, blue or green beams of light, or combinations of the three.

Why is a poinsettia flower red?

In the previous activity you used sweet wrappers as light **filters**. A red filter allows only red light through. The reason that a poinsettia flower is red and a leaf is green is because of a similar colour subtraction system. White light shines onto the flower and the leaf. The **pigments** in the flower absorb all the colours except for the red, which is reflected into our eyes. The pigments in the leaf absorb all the colours and only reflect the green.

In the same way, when white light shines onto a white sheet, all the colours are reflected, and we see it as white. Objects that absorb all the colours and do not reflect any light look black.

Figure 62.3 *A red poinsettia flower absorbs all colours except red.*

ACTIVITY

1. If only green light was shone onto a leaf, what colour would you see it as?
2. Predict what colour a leaf would be under red light only.
3. Test this by covering part of a leaf with a double layer of red Cellophane, and then with green Cellophane.

Figure 62.4 *Our national flag.*

Assessment

Explain why the different parts of the national flag appear as red, black and white.

63

Review topic 4

Light is a form of energy and an electromagnetic wave. It is produced by luminous bodies such as the Sun and travels at high speed in straight lines. When it strikes objects, it bends (is refracted), or bounces back (is reflected); it can also be absorbed. It is a mix of a range of rainbow colours that can be dispersed by a prism.

Light reflected off objects makes it possible for us to see. Our eyes have a lens system that bends the light to focus the image on the retina. Lenses are convex (converging) or concave (diverging) and are for refracting the light in such a way that the image is focused. Lenses are used in optical instruments such as the camera and telescope.

Multiple-choice questions

1 A shadow is formed when light hits a
  a transparent object
  b opaque object
  c translucent object
  d semi-transparent object
2 The splitting of white light into its component colours is called
  a refraction
  b reflection
  c total internal reflection
  d dispersion
3 The principle of the pinhole camera shows that light
  a can be reflected
  b travels in waves
  c travels in straight lines
  d travels fast
4 The iris of the eye is equivalent to the following part of a camera:
  a lens
  b screen with film
  c diaphragm
  d aperture
5 Which of the following statements about a convex lens is incorrect?
  a It is a diverging lens.
  b It is fat in the middle and thin at the ends.
  c It is used to correct long-sightedness.
  d It is used in magnifying glasses.

Longer questions

1 Describe an experiment to show that light travels in straight lines.
2 Are sundials accurate? Explain your answer.
3 a Draw a ray diagram to show how light is refracted by a concave lens.
  b Explain how a lens bends light.
4 a Draw a diagram to show how an image forms in a pinhole camera.
  b Is the image real or virtual?
  c What happens to the image if the size of the hole in the pinhole camera is increased?
5 Draw diagrams showing the eye defects of short-sightedness and long-sightedness and how these can be corrected using lenses.
6 Draw a diagram to show how a spectrum can be formed when white light enters a prism. Label the colours in the correct order.
7 Explain by means of a diagram how you see a leaf as green.
8 a Draw a diagram showing the different structures of the human eye.
  b What are the differences between the human eye and a camera?
9 What colours do you get when mixing the following colours:
  a red and green
  b green and blue
  c blue and red
  d red, blue and green?
10 a What is a mirage?
  b Draw a labelled diagram to show why a straw in a glass of water appears bent.
11 a Draw and label the electromagnetic spectrum.
  b Identify two parts of the electromagnetic spectrum that you use on a daily basis.
  c What are the different wavelengths of white light?
12 a Draw a diagram showing a lunar eclipse.
  b Draw a diagram showing a solar eclipse.
  c Draw a diagram showing the umbra and penumbra of a shadow cast by a ball.
13 Draw rays of light going through:
  a a convex lens
  b a concave lens
  c a prism
  d a rectangular glass block.

Forces and their effects

Keywords

elastic potential energy, elasticity, extension, force, Hooke's Law, kinetic energy

Did you know?

Although they can be easily stretched, materials like chewing gum and Plasticine are not elastic. An elastic substance resists changes to its shape; it goes back to its original shape as soon as the stretching force is removed.

Science Extra

An elastic object will not stretch forever. If you stretch it past a certain point, it will not go back to its original shape. The point is called the *elastic limit* of the material. After this point, the material undergoes plastic deformation.

What is a force?

Whenever you squeeze a lemon, twist open a tap, hit a ball, pull the curtains closed, tear open a packet, write with a pen or type on a keyboard, you are using a force.

We can think of a **force** as a push or a pull, and although we can't see forces, we can see their effects. A force can:

- change an object's motion
- change the shape of an object – temporarily or permanently
- transfer energy.

a) Changing an object's motion.

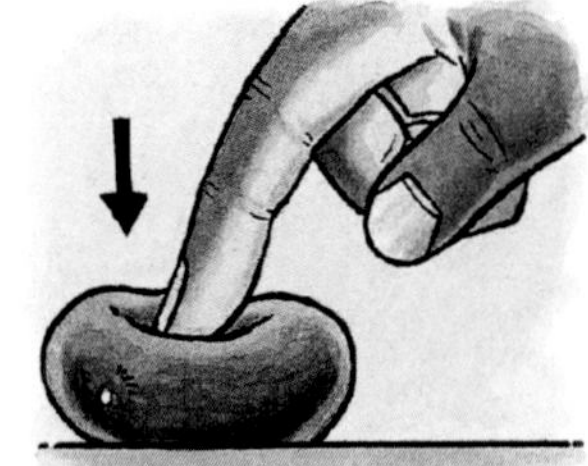

b) Changing the shape of an object.

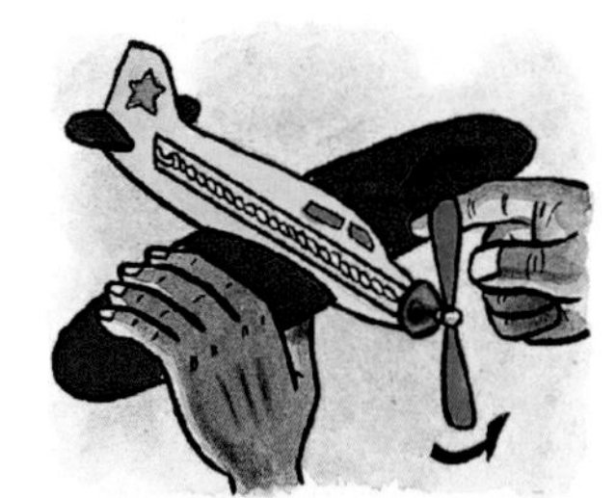

c) Transferring energy.

Figure 64.1 *The effects of forces*

You can see all of these effects by using a catapult in the following activity.

ACTIVITY

Make a catapult using a plastic spoon to shoot a ball of crumpled paper.

1 Put a ball of paper in the cradle of a plastic spoon and hold the spoon by its handle.
2 Apply a force, pulling the cradle of the spoon backwards. Notice how the shape of the spoon changes.
3 Let go of the top of the spoon, and see how far you can fling the paper ball. Notice how the spoon goes back to its original shape.
4 Test how far you can shoot the ball by pulling back on the spoon as far as it can go. What happens if you apply too big a force?

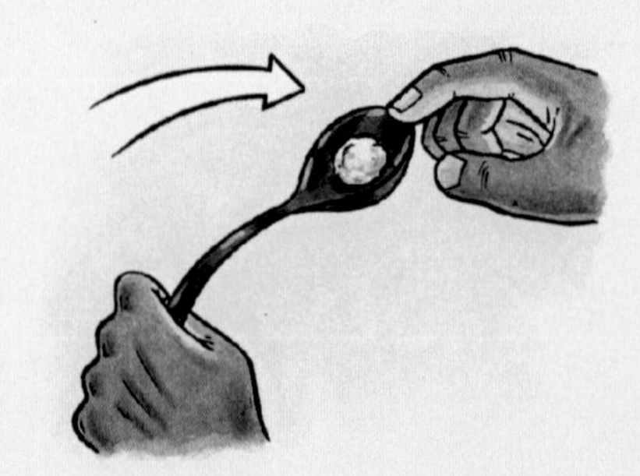

Figure 64.2 *A plastic spoon catapult.*

The force you apply to the spoon temporarily changes its shape. When you let go, the spoon springs back to its original position, and the paper ball is set in motion. The plastic material of the spoon has **elasticity**. When you apply a force to bend back the spoon, energy is stored as **elastic potential energy**. When you let go, this changes into **kinetic** (movement) **energy**. If you apply too big a force to the spoon, it snaps, and you have caused a permanent change in its shape.

While on the subject of elasticity, let's look at Hooke's Law. You saw that to bend the spoon back further, you had to apply a bigger force. **Hooke's Law** states that: The **extension** of a spring is directly proportional to the force acting on it. In other words, how much a spring or elastic material bends or stretches depends on the force that you apply. Weight is a force. (You will see why in Unit 66.)

INVESTIGATION Hooke's Law

If you don't have a spring, use an elastic band and staple it to a piece of card (see the Activity in Unit 66), or hook it around a nail. If you don't have weights, you can use packages of food that have their mass marked on them. A mass of 100 g is equivalent to 1 N.

1 Set up the equipment as shown in the diagram.
2 Copy the results table below into your book.
3 Measure the length of the spring when no weight is added.
4 Add a 0.1 N weight to the spring.
5 Measure the new length of the spring.
6 Add another 0.1 N weight to make a total of 0.2 N, and measure the new length.
7 Continue till you have added 1 N.
8 Did the length of the spring increase by the same amount each time?

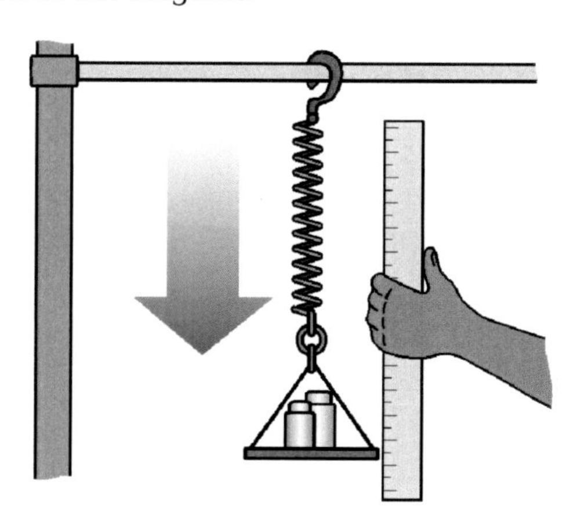

Figure 64.3 *Investigating Hooke's Law.*

Weight (N)	Length (cm)
0.1 N	1 cm
0.2 N	2 cm

If you plot a graph of your results, you should get more or less a straight line.

Assessment

What is a force? Discuss this with a partner for five minutes or less. Then, from what you have learnt so far, write down your own definition of what a force is.

65

Forces come in pairs

Keywords

action, attraction, contact force, electrostatic, interaction, magnetic, non-contact force, reaction, repulsion

Did you know?

Newton wrote three laws about forces that we call Newton's Laws of Motion.

Figure 65.1 *A force is an interaction between two objects.*

Take a look at these pictures of forces at work. Can you see that a force is an **interaction** between two objects? There is an interaction between the ball and the bat, and between the wall and the person leaning against it.

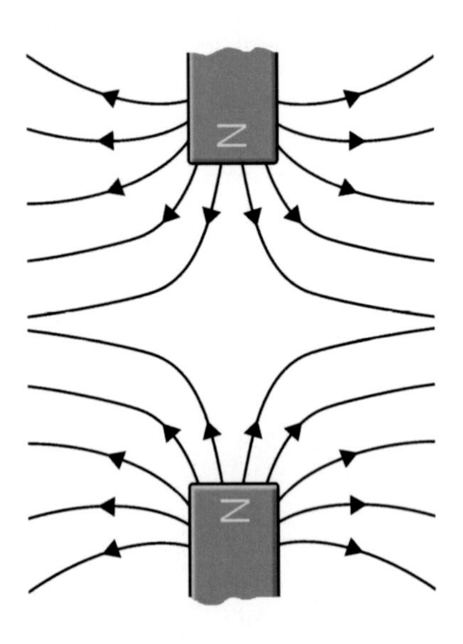

Figure 65.2 *Magnets repelling – a non-contact force.*

Types of forces – contact and non-contact

We can divide forces into two groups, depending on the way the two objects interact. If they interact by touching, like in the examples shown above, **contact forces** are involved. If they interact without touching, the forces are **non-contact forces**, or forces which act at a distance.

Electrostatic and **magnetic forces** are examples of non-contact forces, and they can be forces of **attraction** (pulling towards each other) or **repulsion** (pushing away from each other). Later you will learn more about two different types of forces: friction (a contact force) and gravitational force (a force that acts at a distance). Unlike electrostatic or magnetic forces, gravitational forces are always forces of attraction.

Forces have size (magnitude) and direction

Some forces are big and others are small. We can use arrows in diagrams to show where the forces are acting. In the diagrams at the top of page 135, the direction of the arrow shows the direction of the force, and the length of the arrow shows the size of the force.

Action and reaction – equal and opposite forces

Figure 65.3 *Pushing against a wall gently.*

Figure 65.4 *Pushing against a wall hard.*

Perhaps you are surprised to see in the diagrams that when the person pushes against the wall, the wall pushes back in the opposite direction. Most of our understanding of forces has come from the work of Isaac Newton (1643–1727). One of his observations was that forces act in pairs, and that the two forces are always equal in size, and opposite in direction. This is sometimes called the law of interaction: For every **action** force, there is an equal and opposite **reaction** force. The following activity shows the law of interaction.

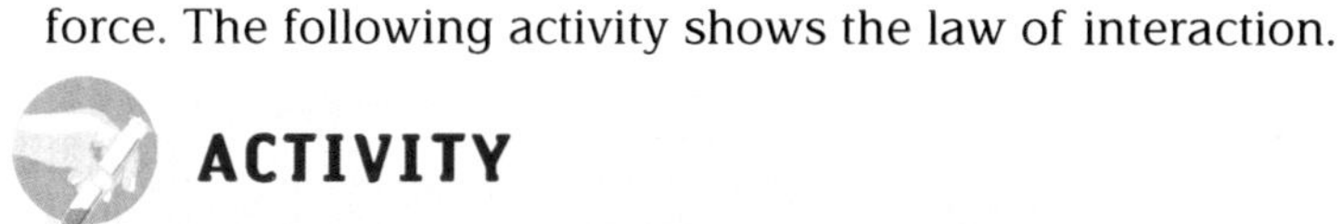

ACTIVITY

1. Pull a long piece of strong string across your desk.
2. Would you agree that you are using a force to pull the string? Would you agree that the string is pulling on you with a force the same size?
3. Now anchor the string to a pole, or a door handle of a closed door. Pull gently on the string. Can you feel the string pulling back on you? When you pull harder on the string, can you feel it pulling back harder?
4. What happens if you pull as hard as you can on the string, standing in your socks on a smooth or slippery floor?

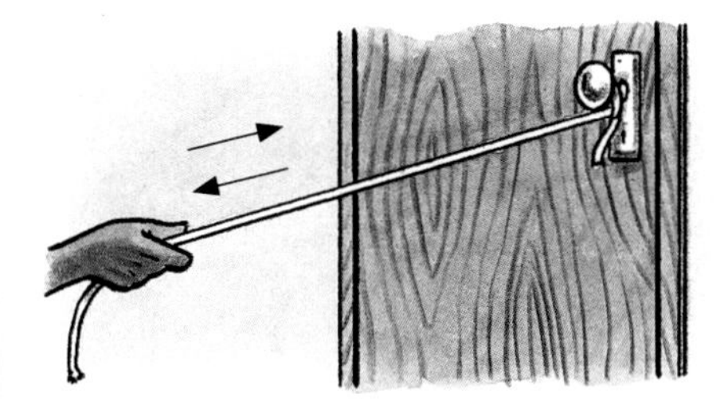

Figure 65.5 *For every action force, there is an equal and opposite reaction force.*

When you pull on the free string, it is hard to believe that it is pulling back on you with an equal and opposite force, but it is. The string moves only because it is much lighter than you are. When you anchor the string, you are putting the mass of Earth behind it, and now you can feel it pulling back.

You can experience this best by pulling hard on the string when there is little friction under your feet. Just as if someone was pulling strongly on the other end of the string in a tug-of-war, your feet begin to slide in the direction that the string is pulling in.

Let's look at the example of a tug-of-war competition. As the diagrams show, all the force pairs are equal, whether the two students pull with the same force or not. As long as the two students pull with the same force, there is no movement. If one student pulls with a bigger force, then it is the overall, or net, force that makes the other student move.

Two students pull with the same force.

Two students pull with different-sized forces.

Figure 65.6 *A tug-of-war competition.*

Assessment

Imagine you have the honour of interviewing Isaac Newton. Working in groups, research one of the following:

- The life of Isaac Newton
- What Newton discovered about forces

and then write a short article, presented as an interview.

66

Measuring force, mass and weight

Keywords

mass, newton (N), newton meter, spring balance, weight

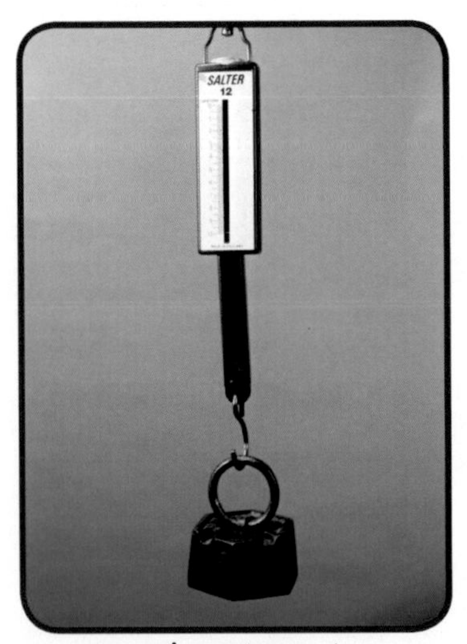

Figure 66.1 *A newton meter.*

Science Extra

Lifting weights is a good example of a pair of equal but opposite forces. Earth pulls downwards on an object with the same-sized force that the bodybuilder must use to lift up the object.

The unit of force is the newton

Forces can be measured because they have magnitude (size). We measure the size of forces using an instrument called a **newton meter** or a **spring balance**. The unit of force, named after Isaac Newton, is the **newton (N)**. You use a force of about 20 N to pull open the tab on a can of cooldrink, so a force of 1 N is quite small. A force of 10 N (more exactly 9.8 N) is needed to lift a mass of 1 kg.

Mass and weight

Remember mass and weight are different. A mass of 1 kg has a weight of about 10 N on Earth. Earth pulls on an object with a gravitational force that we call the **weight** of the object. Like all forces, weight is measured in newtons. (You will learn more about gravitational force in the next unit.)

Mass is the amount of matter in an object, and is measured in kilograms. While mass is constant, weight varies wherever the gravitational force varies. An astronaut who has a mass of 70 kg has different weights on the Moon, Earth, and Jupiter.

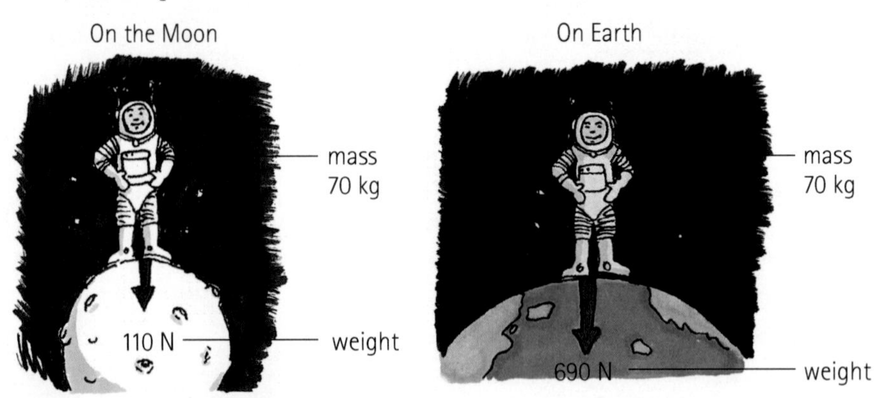

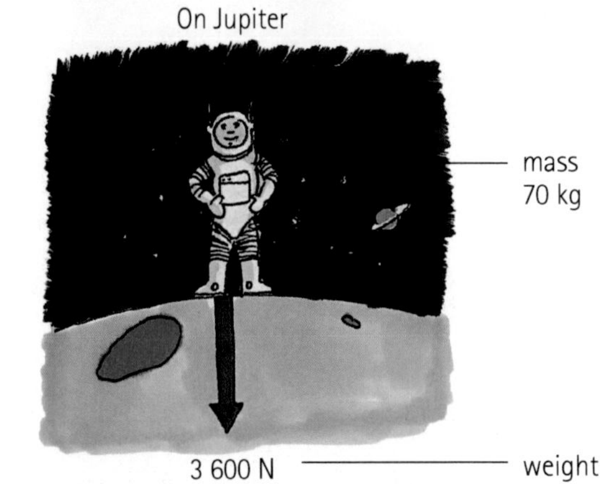

Figure 66.2 *The difference between mass and weight.*

ACTIVITIES

Measuring the size of forces

Use a newton meter or spring balance to measure the size of some forces in the classroom. If you don't have an instrument, make one as shown below.

1 First test the accuracy by measuring the weight of an object of known mass. If you are lifting a mass of 1 kg, you would expect a reading of 9.8 N. If you are lifting a mass of 500 g, you would expect a reading of 4.9 N.
2 Try measuring the size of the force you use for everyday actions. Here are some examples:
   - opening or closing the door
   - pulling the window closed
   - tilting your chair on its back legs
   - lifting an object
   - pulling back the elastic on a catapult.
3 Compare the force you use to lift an object to the force you use to pull the same object across the desk or floor. Can you explain why they are different?
4 Record all your results in a table, like the one below.

Action	Size of force (N)
closing door	11.5
lifting mass of 75 g	

Make a spring balance or newton meter

You can make your own simple spring balance as shown in Figure 66.3 on the right, by stapling a strong elastic band to a piece of card.

You then need to make the scale for your spring balance (and at the same time you can investigate Hooke's Law – see Unit 64).

1 Mark the end of the unstretched elastic band on the card.
2 Place a weight of known mass on the hook – e.g. a 200 g package or 310 ml of cooldrink. Mark the distance that the elastic band has stretched, and label this with the weight in N, e.g. 2 N for 200 g.
3 Repeat with, a 500 g pack of sugar, say, and then 1 000 g and 1 500 g weights.
4 Then use your home-made spring balance to measure some forces in the activity above, using the scale as an approximation. (If you are using this activity to investigate Hooke's Law, measure the distances marked on the card and plot weight (in N) against distance.)

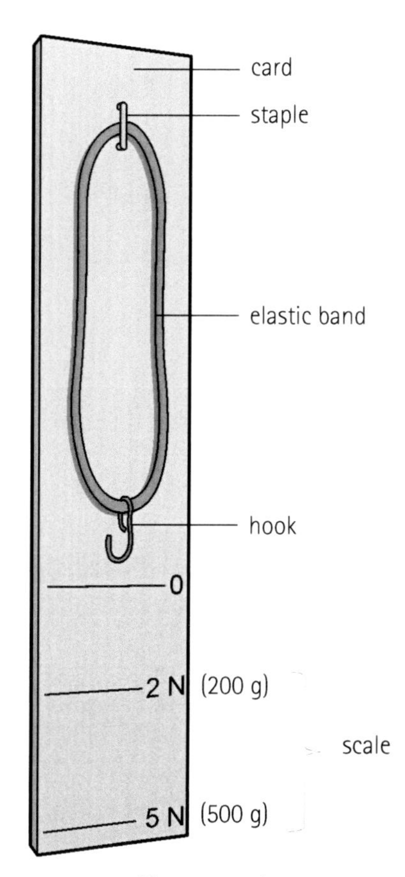

Figure 66.3 *You can make your own spring balance or newton meter.*

Assessment

We often use the term *weight*, when strictly speaking we should say *mass*. Briefly explain the difference between weight and mass. When you lift a weight, what force are you pulling against?

67

Gravity – a force of attraction

Keywords

attraction, *g*, gravity

Did you know?

Newton was the first person to give a scientific explanation of what causes the tides because he discovered the principle of gravity, which includes the Moon's pull on Earth.

Newton and the apple

The story goes that Newton was sitting under an apple tree one day, when an apple fell on him. This jolted him into discovering the principle of **gravity**. Maybe this is just a story, but the important thing is that Newton realised that the same force that makes an apple fall to the ground causes the Moon to orbit Earth.

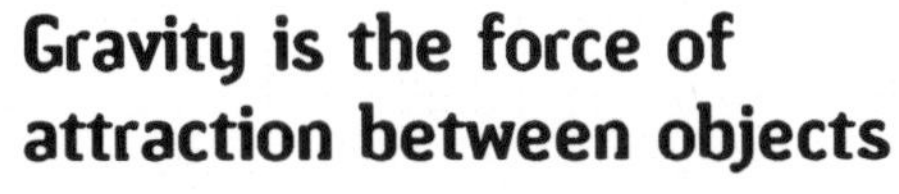

Gravity is the force of attraction between objects

Gravity is a force of **attraction** that exists between all objects. There is a force of attraction between you and the person sitting next to you. There is a force of attraction between the desk and the chair. All masses attract each other.

The strength of this force of attraction depends on the masses of the two objects and the distance between them. Two 1 kg masses placed 1 m apart have a force of attraction of only 6.7×10^{-11} N between them. But because Earth has a huge mass (6×10^{24} kg), its pull is very strong.

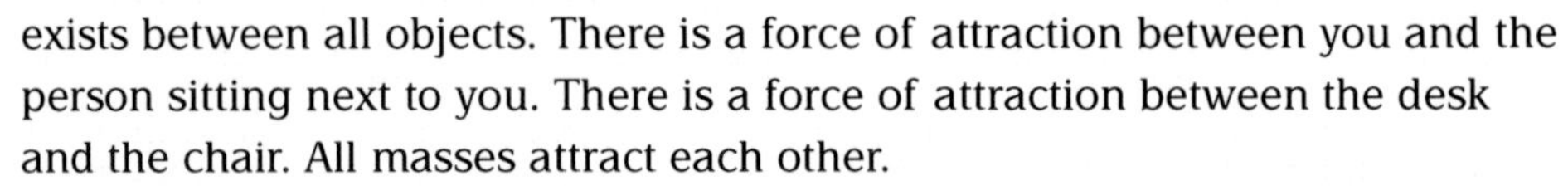

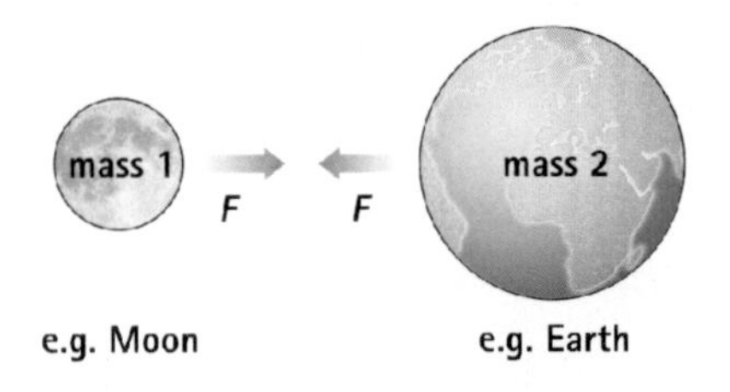

Figure 67.1 *Gravity depends on the masses of two objects and the distance between them.*

Figure 67.2 *A cyclist speeds up quickly downhill because of the force of gravity.*

The pull of Earth accelerates falling objects

Gravity, like all forces, can change the state of motion of an object. If you throw a ball high up in the air, it doesn't keep on going up. It slows down, hangs in the air for a moment, and then falls towards the ground, speeding up as it goes. Gravity makes falling objects accelerate. We call the acceleration due to gravity, ***g***, and it has a value of 9.81 m/s^2 . This means that all objects should fall at the same rate, which they would if there were no air resistance.

force = mass × acceleration

Can you now see why a 1 kg mass has a weight of 9.8 N?

According to legend, Galileo dropped balls of different masses from the top of the Leaning Tower of Pisa to prove his theory that they should fall at the same rate. He reasoned that although a bigger object would have more gravitational force, it would also have more inertia. (You will learn about inertia later.)

Figure 67.3 *The famous Leaning Tower of Pisa in Italy.*

ACTIVITIES

1 A feather and a hammer would fall at the same rate if there were no air resistance. Go on the Internet and watch a video of this experiment being done on the Moon. Here's an address for you to try: http://video.google.com/videoplay?docid=6926891572259784994.

2 Read 'Gravity is less on the Moon' below and then work out the value for acceleration due to gravity on the Moon.

Gravity is less on the Moon

Since the Moon is so much smaller than the Earth, the Moon's gravity **is only** $\frac{1}{6}$ of Earth's. Not only do objects weigh six times less on the Moon, **but** when objects are dropped on the Moon, they fall six times more slowly. If you drop an eraser from a height of about 1 m, it takes about half a second to hit the floor. On the Moon, it would take three seconds.

Science Extra

The force of gravity depends on the masses of the two objects, and the distance between them, from their centres. The mass of the Moon is actually only about $\frac{1}{80}$ the mass of the Earth, but because the Moon is also much smaller, astronauts on the surface are much closer to the centre of the Moon. This is why your weight on the Moon would be $\frac{1}{6}$ of your weight on the Earth.

Assessment

A body or object has weight when it is in a gravitational field. Imagine you had an anti-gravity spray that cancelled out the gravitational force exerted by Earth. Then imagine that you sprayed it around in your home or the classroom. Write a short story about the effects. (You can read about weightlessness in Unit 47 of Topic 3 to help you.)

68

Friction – a force that resists movement

Keywords

air resistance, aquaplaning, friction

Did you know?

The non-stick Teflon coating of many frying pans was developed by the US space agency, NASA. Teflon makes a very smooth surface on the outside of spacecraft, and so reduces the friction with the atmosphere.

Friction is a force between surfaces in contact

Friction is a force between two surfaces that rub against each other. It's a force that resists a sliding motion. If, for example, the engine of a moving car on a flat road is switched off, the car slows down and eventually comes to a stop. This is because of:

1 friction between the tyres of the car and the road, and
2 the resisting force of the air that the car is moving through, i.e. **air resistance**.

Friction is a force that acts in the opposite direction to movement.

Useful and unwanted friction

Wheels were an important invention that reduced friction – imagine pushing a trolley or a car without them. But wheels also rely on friction. If there was no friction between the tyres of a car and the road, for example, the wheels would spin, but the car wouldn't move. Think of a car wheel stuck in the mud.

Figure 68.1 *Wheels help to reduce friction!*

Friction is also needed to slow or stop a car – it's the friction between the brake pad and the disc that slows the turning wheel down. And so friction is important for road safety. If, when a car brakes, the road surface or the tyres don't offer enough friction, the car will skid. **Aquaplaning** is an example, when the road is wet and slippery. The tyres no longer make contact with the road surface, and the car skids on a thin film of water.

Friction also keeps you on your feet – most of the time! Whenever you slip, it is on surfaces that are wet and smooth, or in shoes with soles that don't have enough grip.

Figure 68.2 *The tread on a walking boot gives good grip.*

Figure 68.3 *Olympic swimmers shave their legs and arms to reduce friction, making their bodies as streamlined as possible.*

Science Extra

Friction between two surfaces generates heat. Kinetic (moving) energy is changed into heat energy. We make use of kinetic energy when we rub our hands together to warm them up.

ACTIVITY

1 Wherever there is movement, there is friction. Brainstorm some of the places where friction is important.

Then

2 Keep a friction diary for a day.

OR

Give examples of where friction operates in the games that you play.

INVESTIGATION Which surfaces have the highest friction?

You can probably predict that rough surfaces increase friction, and smooth ones reduce friction. This is not always the case, though. Rubber can be smooth but it's soft and 'sticky', and therefore offers quite a bit of friction. Test some surfaces by doing the following.

Figure 68.4 *Investigating friction.*

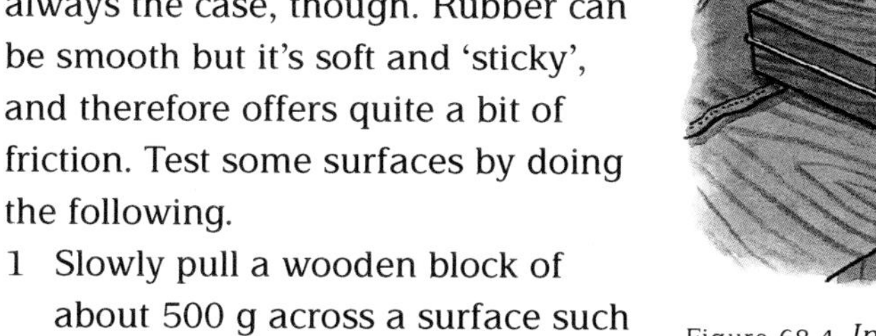

1 Slowly pull a wooden block of about 500 g across a surface such as wood or Melamine, using a newton meter. Record the size of the force you needed to apply.
2 Repeat this experiment on different surfaces, or on the same surface that you now treat by polishing it, or by coating it with oil.
3 Also test what happens if your partner pushes down on the block, perpendicular to the direction you are pulling in. Explain why this increases the friction.
4 Present your results in a table, like the one shown below, giving a friction rating for each surface, according to your measurements.

Surface	Force applied to move block (N)	Friction rating
Melamine		
Melamine lubricated with oil		
Sandpaper		

Assessment

Look at the pictures A, B and C in the margin, showing examples of friction. For each, indicate:

- where the friction is taking place
- whether it is useful or unwanted friction and, depending on your answer, determine how it can be increased or decreased.

A

B

C

Figure 68.5 *Examples of friction.*

69

Movement without force – it just keeps on going

Keywords

motion

Did you know?

It was the Italian, Galileo, who suggested that it is friction that brings moving objects to a stop. Newton was born the year Galileo died (1643), and developed Galileo's thinking further.

A force makes a moving object come to a stop

You now know that when a moving car's engine is switched off, it is friction that eventually brings the car to a stop. And when a cricket ball flies through the air, it is the force of a fielder's hand that stops its movement or, if the ball is not caught, it is the force of friction. First air resistance slows the ball down, and once it hits the ground, friction between its surface and the grass eventually stops it.

You can now see that it is a force that makes a moving object come to a stop.

Figure 69.1 *The friction between the surface of the cricket ball and the grass eventually stops the ball from moving further.*

Movement without force

A moving object keeps moving unless a force acts on it. In other words, if it weren't for friction, a moving object would keep moving. Newton explained it like this: Imagine a ball on a frictionless track like the one shown in the diagram below.

With the force of gravity acting on it, the ball moves up and down between the two slopes forever. If one side of the track were made flat, then the ball would keep going in a straight line along the track without ever stopping. If there is no force acting on the ball, there is nothing to slow it down.

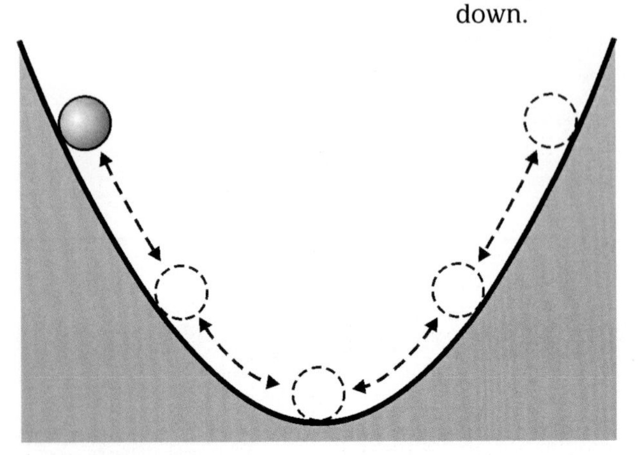

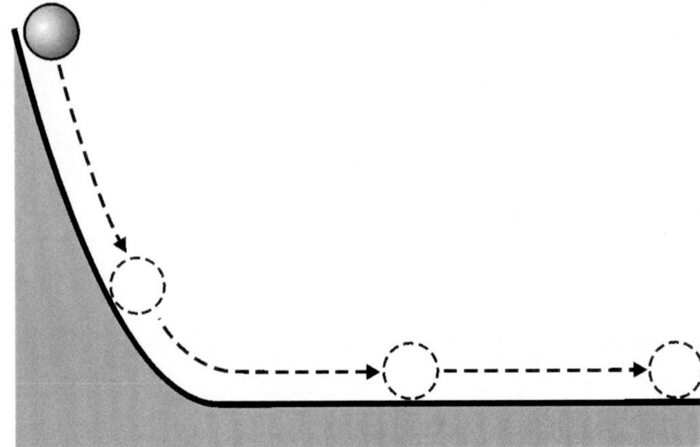

Figure 69.2 *Non-stop motion: A ball rolling on a frictionless track would never slow down or come to a stop.*

So, surprising as it may seem, a force is not needed to keep an object moving at a steady speed. The best place to test this is in space, where there is no air friction. Think of a satellite or space shuttle orbiting the Earth. Once a spacecraft has been set into orbit at a speed that counters the pull of the Earth, no force is needed to keep it moving.

Figure 69.3 *A space shuttle orbiting Earth moves at a steady speed with no force acting on it.*

ACTIVITY

For the space shuttle shown in the diagram above, discuss what is needed:

- to speed it up
- to change its direction.

The First Law of Motion

Newton's First Law of Motion tells us that, in the absence of a force, 'a moving object will keep moving at the same speed, and in the same direction. Without a force acting on it, a stationary object will stay where it is.' So we can add to our definition of a force: A force is anything which changes or tends to change the state of rest or uniform **motion** of a body in a straight line.

Assessment

More than 2 500 years ago, Aristotle (the Greek scientist and philosopher) proposed that a moving object will always bring itself to a stop. Explain why he came to this conclusion and also why it is incorrect.

70

Inertia – why you must wear a seat belt!

Keywords

inertia

Did you know?

Inertia comes from the Latin word meaning *laziness.*

Science Extra

If you have access to the Internet, have a look at the animation of 'The Truck and Ladder' at http://www.geocities.com/Athens/Academy/9208/il.html.

Inertia on a bus

Imagine you get onto a crowded bus and there's only standing room. The driver lets out the clutch too fast, and the bus shoots forwards. You and the other standing passengers are thrown to the back of the bus. That is how you would describe what happens, but in reality, because the bus moves forwards so suddenly, you are 'left behind'.

Figure 70.1 *Inertia on a bus: reluctant to start.*

No sooner have you recovered from this, than the driver brakes without warning while the bus is speeding along. This time, you are all thrown forwards; you keep moving, even though the bus has stopped.

Figure 70.2 *Inertia on a bus: reluctant to stop.*

What you have experienced on this unsettling bus ride is **inertia**: All bodies resist being set in motion, and once they're moving, they're difficult to stop. Worded slightly differently, we can say: 'An object in motion wants to stay in motion, and an object at rest wants to stay at rest.'

Here are two simple activities to demonstrate inertia.

ACTIVITIES

1 **Reluctant to start**

Place a glass, half-filled with water, on a piece of paper on the edge of a table. To pull the paper out from under the glass: Hold the end of the paper sticking out beyond the table, and using a ruler like a knife, hit the paper hard.

What happens to the glass of water? Explain.

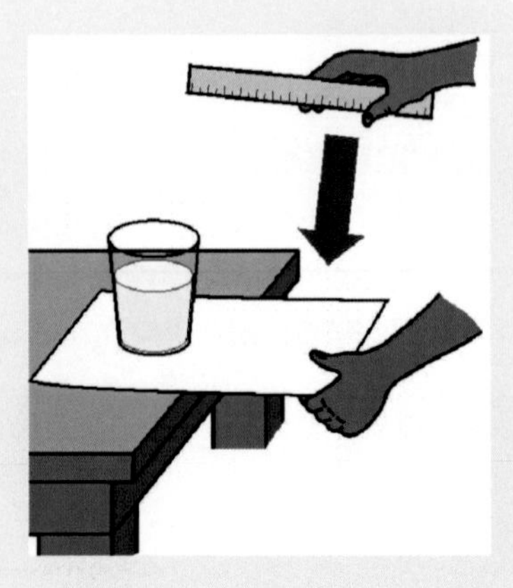

2 **Reluctant to stop**
Place a ball or an orange at one end of an open shoebox.
Push the box along the floor, so that the front end of the box hits the wall.
What happens to the ball or orange? Explain.

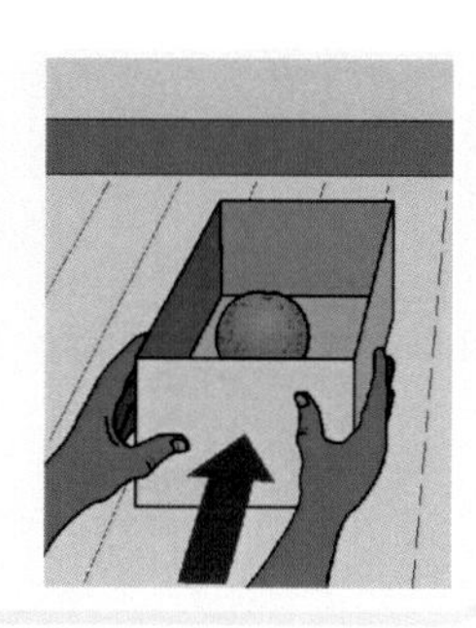

Mass is a property of inertia

The heavier an object is, the more inertia it has, and the more difficult it is to speed up, slow down, or change direction. You can demonstrate this by repeating the activity with the orange in the shoebox. This time add a marble to the box, and give the box a gentle push. You can also do the following investigation.

INVESTIGATION How does mass affect inertia?

Method A
Set up your investigation as shown in Figure 70.3.

1 Place two blocks of different masses on ball bearings or marbles in a long tray. (The ball bearings reduce friction.) Or you can use two identical trolleys, loaded with different masses.
2 Add small equal weights to the end of a string attached to each mass.
3 Let go of the weights at the same time. What happens to the two masses? Record the time it takes for each block or trolley to reach the edge of the table. Which one speeds up more quickly? Which one resists the change to its state of rest the most, i.e. which one has more inertia?
4 If both of the blocks were moving at full speed, which one would be more difficult to slow down, i.e. which one has more inertia?

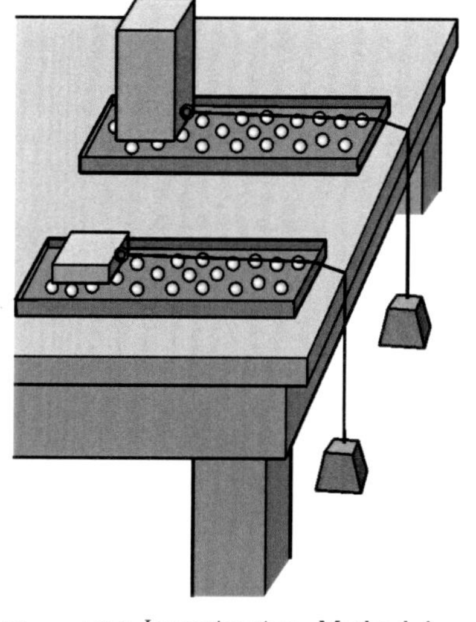

Figure 70.3 *Investigation, Method A.*

Method B

1 Hang two identical tin cans – one empty and one filled with wet sand – from a rail or coat hanger. Use strings of the same length.
2 Push each tin lightly to make them swing. Compare how much harder you have to push the heavier tin to get it to swing to the same height as the empty one. Why is this?
3 Once the cans are swinging at the same height, which one is easier to stop? Why is this?

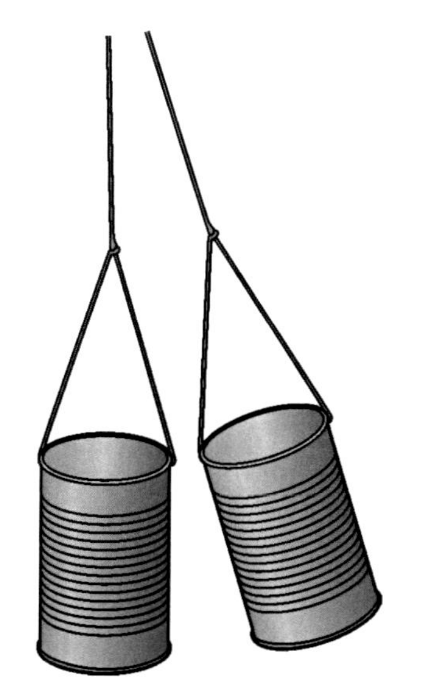

Figure 70.4 *Investigation, Method B.*

Assessment

If you are being charged by a rhinoceros, the advice is to wait until it is almost upon you, and then to dodge to the side. Use the principle of inertia to explain why this is sensible advice.

71

Momentum – a combination of mass and speed

Keywords

collision, momentum

Did you know?

Cars are designed to crumple when they crash. The momentum of the moving car is absorbed into the crumpling of the metal and so protects the passengers.

Science Extra

What is the difference between speed and velocity? Speed is a scalar quantity, which means that it only has size. Velocity is a vector quantity, which means that it has both size and direction.

Mass, speed, momentum

You are walking in a narrow corridor at school and two boys, running at full speed, are coming towards you. One is big and the other is small. You can't avoid a **collision**, but which boy would you rather collide with? In another instance, two girls of the same size are moving towards you. One is walking, the other is running. Again, you can't avoid a collision, so which one would you prefer to bump into?

You know from experience that you would rather collide with the small boy than the large one, and with the girl who is walking, not running.

Figure 71.1 *Momentum depends on mass and speed.*

In the first example, we had two objects (boys) with different masses, moving at the same speed. In the second example, we had two objects (girls) with the same mass, moving at different speeds. **Momentum** is a measurement of motion, and it depends on the mass of the object and its speed. The word *momentum* comes from Latin, and means *movement* or *moving power*.

momentum = mass (kg) × velocity (m/s)

Using the formula above, we can work out the momentum of the two boys running at the same speed (5 m/s) as follows:

If boy 1 has a mass of 45 kg and boy 2 has a mass of 70 kg, then:
Boy 1's momentum = 45 kg × 5 m/s = 225 kg m/s
Boy 2's momentum = 70 kg × 5 m/s = 350 kg m/s

Now you might be asking, 'If a moving object has inertia (the reluctance to stop), and a moving object has momentum, what is the difference between inertia and momentum?' Unlike inertia, momentum is a measure of motion. A body at rest has inertia, but no momentum. A moving body has both inertia and momentum.

ACTIVITY

1 A car has a mass of 1 200 kg and travels at 40 m/s. A truck has a mass of 6 200 kg and travels at 25 m/s. Which one has the bigger momentum? Use your answer to explain why trucks sometimes have lower speed limits than cars on a highway.
2 A stationary car has a speed of 0 m/s and a mass of 1 200 kg. Does it have (a) momentum and (b) inertia?
3 A bowler who can deliver a ball at speeds of over 140 km/h qualifies as a fast bowler. Here are some ball speeds recorded for international matches.

Bowler	Team	km/h	Against	Date
Shoaib Akhtar	Pakistan	161	New Zealand	27/04/2002
Brett Lee	Australia	157.4	South Africa	08/03/2002
Franklyn Rose	West Indies	146.4	England	26/06/2000

A cricket ball has a mass of only 0.15 kg, but show, using Franklyn Rose's bowling speed record, how Rose can give a ball a lot of momentum.

Figure 71.2 *Newton's cradle.*

Transfer of momentum

Can you predict what happens when you set one of the balls of a Newton's cradle in motion? A special property of momentum is that it can be transferred. Imagine a head-on collision between a large truck and a small car, both moving at the same slow speed. As the truck has greater momentum, it will force the car to move backwards.

Figure 71.3 *When this car and truck collide, the truck transfers its momentum to the car.*

ACTIVITIES

Try all or some of these to demonstrate the transfer of momentum.

1 Make a track for marbles on a flat surface between two rulers. Place one marble in the middle of the track and roll another marble into it. Observe what happens. Test what happens when you increase the number of stationary marbles you place in the middle of the track, and/or the number of marbles you shoot.
2 If you have a Newton's cradle, test what happens when you raise the ball to different heights, and the number of balls you lift. Or you can try out the interactive simulation on http://www.school-for-champions.com/science/newtons_cradle.htm.

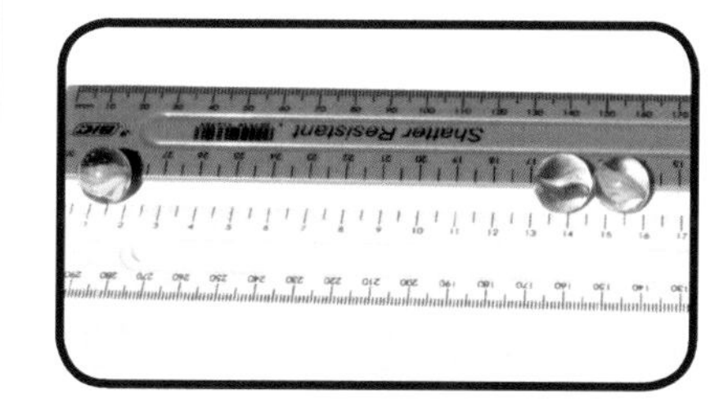

Figure 71.4 *Demonstrating the transfer of momentum.*

Assessment

Is momentum a force? If your answer is no, explain what it is.

72

Acceleration – the need for speed

Keywords

accelerate, acceleration, decelerate

Did you know?

We use the term deceleration to make it clear that it is different from an increase in speed. But scientifically, the term acceleration refers to any change in speed over time, whether up or down.

Science Extra

A force of 1 N is the force required to accelerate a mass of 1 kg at 1 m/s^2.

Force, mass and acceleration

A karate master uses the whole mass of his body (e.g. 60 kg) and a high acceleration of the hand and forearm (e.g. 50 m/s^2) to produce a force big enough to break these planks in half. When a force acts on an object, the strength of the force = mass × acceleration. If we use this formula, we can calculate that the force used to break the planks is:

Figure 72.1 *The secret of the karate chop is in the acceleration.*

60 × 50 = 3000 N

Compare this to the weight of the karate master, which is 600 N.

The following investigation demonstrates the relationship between force, mass and acceleration.

INVESTIGATION How are force, mass and acceleration related?

Part A: The effect of mass on force

1 Drop two objects of different masses (but the same size and shape) into a tray of sand or flour. What is the acceleration due to gravity for each of the objects?

2 Observe the size of the dent or crater formed in the sand. How can you relate this to the force of impact?

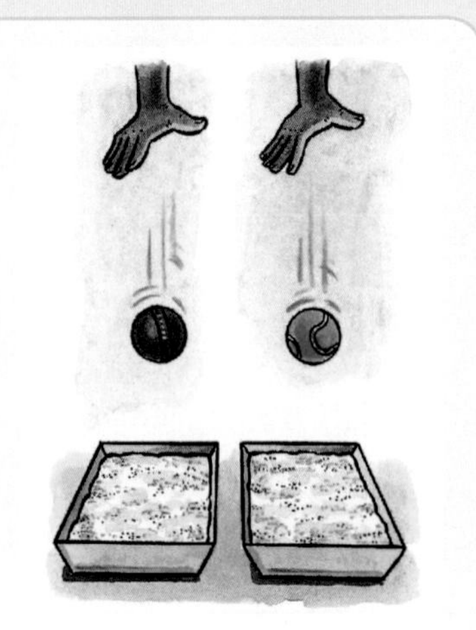

Figure 72.2 *Testing the effect of mass on force.*

In part A of the investigation, you are comparing different masses, while acceleration is constant. The acceleration due to gravity of the two falling objects is the same (9.8 m/s^2), but the force of impact depends on the mass of the object. The greater the mass, the greater the force.

INVESTIGATION How are force, mass and acceleration related?

Part B: The effect of a force on acceleration

1. Set up the experiment as in Figure 72.3. Attach a small weight on the end of the string to a trolley, toy car or roller skate to act as a force.
2. Measure (with a stopwatch, or by counting) the time it takes for the trolley to move across the table.
3. Repeat the experiment using a heavier weight to produce a larger force. You can assume that the time is directly proportional to the acceleration.
4. What happens to the acceleration of the trolley as you increase the force (i.e. the weight)?

(Alternatively, you can push two roller skates or toy cars of the same mass, across the floor to a finish line, giving one a harder push.)

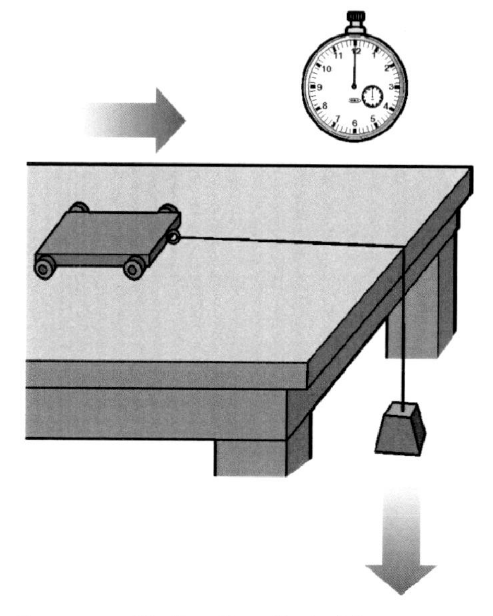

Figure 72.3 *Testing the effect of force on acceleration.*

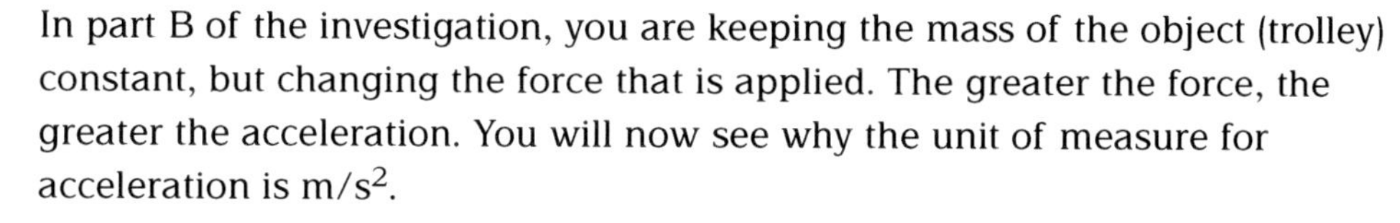

In part B of the investigation, you are keeping the mass of the object (trolley) constant, but changing the force that is applied. The greater the force, the greater the acceleration. You will now see why the unit of measure for acceleration is m/s^2.

Acceleration and deceleration – forces produce a change in speed

How fast an object can speed up (**accelerate**) or slow down (**decelerate**) depends on its mass, and the force applied. This Formula One racing car has a light frame, a powerful engine and a remarkable braking system for very quick changes in speed. The car can go from 0 to 160 km/h and back to 0 again in less than five seconds.

Acceleration is the increase in speed over time. The car can accelerate from 0 to 100 km/h in 1.9 seconds. We can work out the acceleration rate for the car as follows:

Figure 72.4 *This Formula One racing car can accelerate from 0 to 100 km/h in less than 2 seconds.*

$$\text{Acceleration} = \frac{\text{speed 2} - \text{speed 1}}{\text{time}}$$

$$= \frac{100 \text{ km/h} - 0 \text{ km/h}}{1.9 \text{ s}}$$

$$= \frac{52.6 \text{ km/h}}{\text{s}}$$

And if we convert km/h to m/s:

$$= \frac{52\,600 \text{ m}/60 \times 60 \text{ s}}{\text{s}}$$

$$= 14.6 \text{ m/s}^2$$

The Formula One car accelerates at 14.6 m/s^2. This is more than 1.4 g or 1.4 times the acceleration due to gravity. This means that as the car accelerates, the driver is pushed back in his seat with a force that is 1.4 times his body weight.

Assessment

If a racing car driver's mass is 70 kg, and the car he's driving is accelerating at 14 m/s^2, calculate with what force he is pushed back in his seat. Normally he weighs 700 N. Can you see how this is an example of how acceleration can affect the size of a force?

73

Forces in bobsledding

Keywords
aerodynamic, centrifugal force, sliding friction, static friction

Did you know?

If a bobsled reaches the finish line without all of its team members in the sled, the team is disqualified. A sled is allowed to cross the line on its side or upside down, as long as everyone's inside it.

Jamaica has a team that competes in the bobsled event at the Winter Olympics. A bobsled has no engine, but runs on a track of ice. The course has many bends and turns, and is downhill most of the way. Push and gravity are the only sources of speed. The sled must make use of those forces that help it accelerate, and minimise the forces that slow it down. Let's look at the forces that a bobsled team has to work with or against.

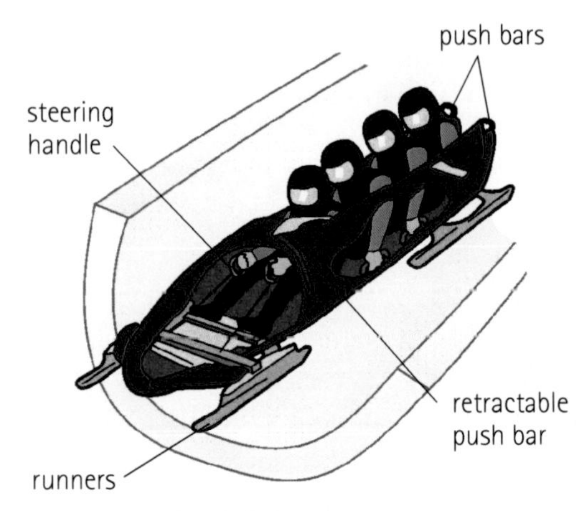

Figure 73.1 *A bobsled with its team.*

Friction and air resistance

As the track is ice, friction between the metal runners of the sled and the ice is low. The use of lubricant isn't allowed, but the metal runners are made as thin and as smooth as possible. The surface of the ice roughens with each run, due to scratching and scarring, so friction can increase on the track during the race.

Air resistance or drag during the ride is the biggest source of friction, and the sled is designed to be as streamlined as possible. The push bars, for example, are collapsible. Also, the team members wear skin-tight, **aerodynamic** suits.

Gravity and momentum

Acceleration due to gravity is 9.8 m/s^2 and momentum is mass × speed. The heavier an object, the faster it can go with gravity's help – but only once it is going. There is a maximum mass allowed for the bobsled and team. If a team doesn't reach the maximum occupied mass, they'll often add ballast (extra mass) to improve momentum. But the heavier the bobsled, the harder it is to give it a fast push-off at the start of the race.

The push-off

The race begins with a stationary bobsled. The team has to get the bobsled going over a distance of 50 m. It's hard work because of inertia, and because the **static friction** (friction between the stationary sled and the ice it is resting on) is greater than the **sliding friction** (friction between the moving sled and the ice).

Figure 73.2 *The bobsled push-off: static friction is greater than sliding friction.*

For this reason, the members of a team have to be strong athletes. Steering the bobsled is also hard work, and so, to save his or her strength, the driver on the team usually only helps to push once the sled is moving. The team members wear shoes with brushes to give them grip or traction on the ice. During the push-off, the team accelerates the sled from 0 to 40 km/h.

Science Extra

During the push-off, the bobsled team has to get the sled going and they need to be strong athletes to do this work. In physics terms, work = force × distance.

The ride

Once the sled is moving fast enough, the push bars are folded in. The team jumps into the sled (the braker is usually the last to get in) and they crouch low to reduce air resistance. Gravity and momentum take over, and the race is now up to the driver, whose job it is to steer the sled. The rest of the team members are there to contribute to momentum.

The driver aims for the best possible path on the bends. Too low, and the sled does not take advantage of the **centrifugal force** (the force experienced by an object moving in a circular path). Too high, and the sled travels over an unnecessarily longer distance. The sled reaches a speed of up to 130 km/h. Crashes happen quite often. The brake is usually only applied once the bobsled has crossed the finish line.

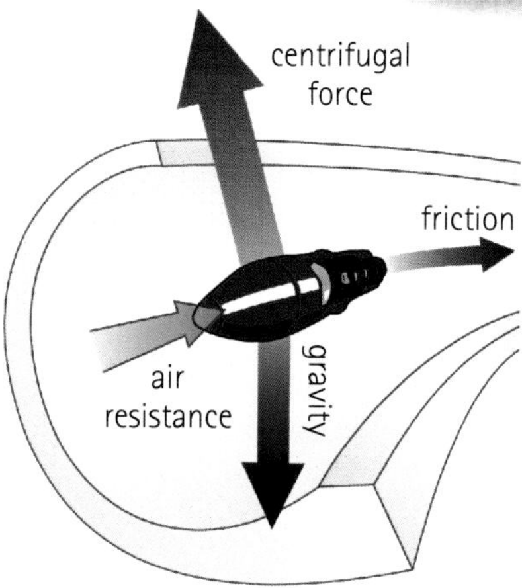

Figure 73.3 *Some forces acting on a moving bobsled.*

ACTIVITY

1. Name a force that slows the bobsled down, and a force that speeds the bobsled up.
2. What is the property that explains why it is hard work to get the bobsled going and then hard work (a strong braking force is needed) to get the bobsled to stop?
3. If the masses of the team members are 64 kg, 71 kg, 75 kg and 77 kg and the mass of the bobsled is 200 kg, calculate the momentum of the bobsled:
   a. just after push-off, at 40 km/h
   b. at a full-speed of 130 km/h.

   Why do you think there is a maximum allowed mass for the bobsled and the team?
4. Calculate the acceleration of the bobsled during the push-off if it goes from 0 to 40 km/h in 6 seconds.

Assessment

The study of forces is a key part of physical science. You have seen some everyday examples of where forces apply. Now that you have learnt all about forces, write a one-page essay, describing what a force is. Compare this with the answer you gave in Unit 64.

Pressure – the force on a surface

Keywords
area, pascal (Pa), pressure

Did you know?
The unit of pressure is named after Blaise Pascal – the French scientist, mathematician and philosopher who lived in the 1600s.

When you push a drawing pin into a piece of wood, the force from your finger is concentrated on the sharp point (a very small area), producing a high pressure. Pressure tells us how concentrated a force is.

Units of pressure

We define **pressure** as the force acting over a given **area**. The unit of pressure N/m^2 is also called the **pascal (Pa)**.

$$\text{pressure} = \frac{\text{force}}{\text{area}} = \frac{N}{m^2}$$

INVESTIGATION How does area affect pressure?

1. Spread a 2 cm thick layer of Plasticine on a small piece of cardboard.
2. Put this on a scale and set the scale to zero.
3. Place a block with 1 cm sides on top of the Plasticine, and push down on the block until the reading on the scale is 1 N (100 g).
4. Repeat this, using a block with 2 cm sides.
5. Answer the following:
   a. Which block makes a deeper dent in the Plasticine?
   b. What is the area of one side or face of the 1 cm block?
   c. What is the area of one side of the 2 cm block?
   d. Use the formula above to calculate the pressure for each block in N/cm^2.
   e. How many times less is the pressure for the 2 cm block?

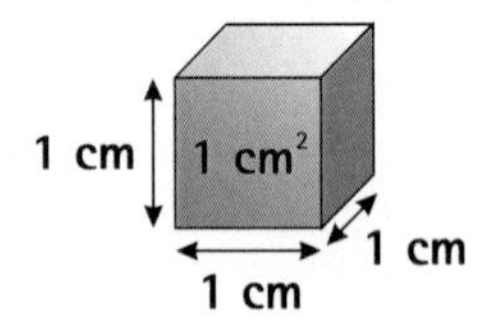

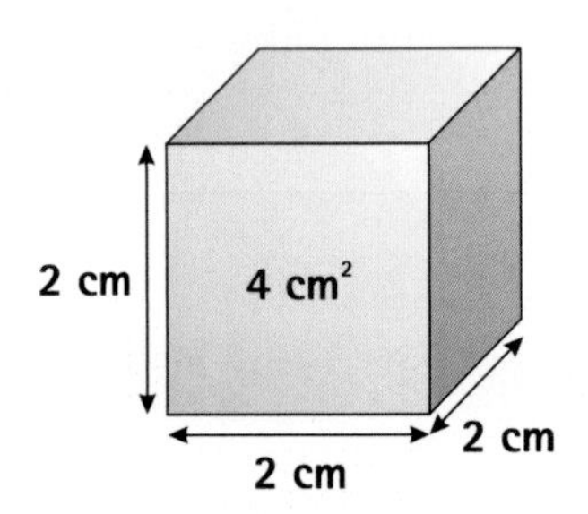

Figure 74.1 *The areas of a 1 cm and a 2 cm block.*

The relationship between pressure, force and area

In the investigation above, you will have seen that although you pressed with the same force (1 N), the pressure dent made in the Plasticine by the larger block is not as great as the one made by the smaller block. This is because the larger block has a larger area (4 cm^2), which spreads out the force you apply.

Using the formula for pressure, you can calculate the pressure for the two different blocks:

Pressure for the 1 cm block $= 1\ N/1\ cm^2 = 1\ N/cm^2$
Pressure for the 2 cm block $= 1\ N/4\ cm^2 = 0.25\ N/cm^2$

In the same way as you can increase pressure by applying a force to a small area (think of the drawing pin), so you can reduce pressure by spreading a force over a larger area. This is why camels have wide feet:
to spread their weight over a larger area, which reduces their pressure on the ground and stops them from sinking too easily into the loose sand as they walk.

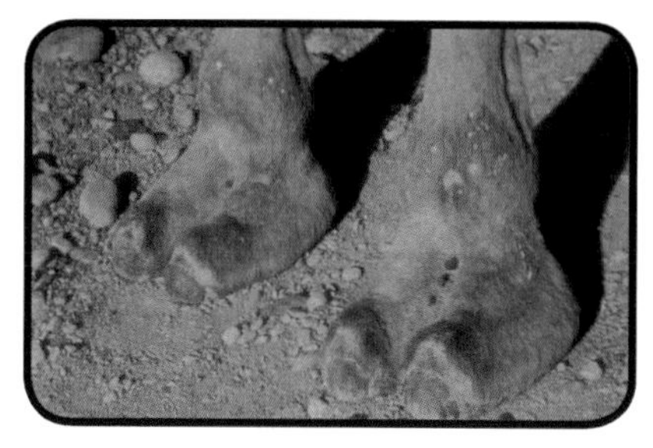

Figure 74.2 *A camel's feet are adapted to their sandy habitat.*

ACTIVITY

Work out how much pressure you exert on the ground when you are standing. Do this by weighing yourself and measuring the area of your feet as follows.

1 Weigh yourself, giving the answer in newtons.
2 To work out the area of your feet, stand on a sheet of squared paper (you can make your own with 2 × 2 cm squares), and draw around your feet.
3 Then count the number of squares that were covered by your feet. (You can do this for one foot and then multiply by 2, or measure both feet and compare their areas.)
4 Calculate your pressure in N/cm^2 when you stand on two feet.
5 How much pressure would you exert if you stood on one foot?
6 Convert your answers to N/m^2. [Hint: $1 m^2 = 100 cm \times 100 cm = 10\,000 cm^2$.]

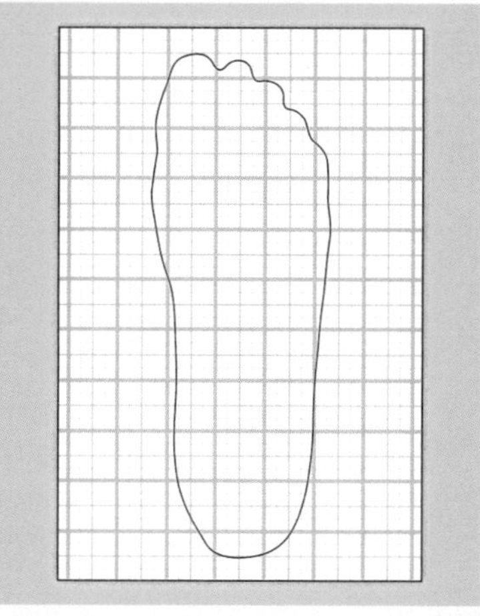

Figure 74.3 *Outline of foot on graph paper.*

How to calculate your pressure:
Number of complete squares = 22
Incomplete squares add up to 13
Total number of squares = 35
Each square = 2 cm × 2 cm
= $4 cm^2$
Total area = $35 \times 4 cm^2$
= $140 cm^2$

Science Extra

Spreading weight over a big area makes it possible to lie on a bed of nails without injuries or punctures. But standing barefoot on the bed of nails is another matter! Now that you know about pressure, you can understand why.

Assessment

Look at the picture in the margin. Explain why it might be less painful if an elephant stood on your foot than if a woman wearing shoes with stiletto heels did so.

75

Atmospheric pressure

Keywords

aneroid barometer, atmospheric pressure, barometer

Did you know?

Atmospheric pressure at sea level is about 100 kPa. This is equal to the weight of ten cars pressing down on one square metre.

Air has weight and even though you don't feel it, there is a lot of it pressing down on you. This pressure from the air is called **atmospheric pressure** or air pressure, and it's enough to crush a can or hold a piece of paper in place under an upturned mug of water.

ACTIVITIES

1 Upside-down mug of water

- Fill a mug or glass to the brim with water.
- Place a piece of paper or card over the mug and press it against the rim.
- Holding the card tightly in place, carefully turn the mug upside-down.
- Let go of the card.

2 Collapsing can

- Pour a little water into a metal can that has a tight-fitting lid.
- Without the lid on, heat the water in the can until it is boiling strongly.
- Remove the can from the heat and immediately put the lid on it tightly. Watch what happens.

3 Drinking straw

- Use a small nail to make a hole in the lid of a plastic bottle, just big enough to fit in a straw.
- Fill the bottle with water, screw on the lid and put a straw through the hole.
- Make the hole airtight with sticky putty or Plasticine.
- Try taking a sip of water through the straw.

4 Plunger suction

- Push two household plungers together to squeeze out all the air.
- Try to separate them again just by pulling.
- Can you explain what is holding the two plungers together? (Hint: Read the explanation on the next page about the tin can and pressure differences.)

1

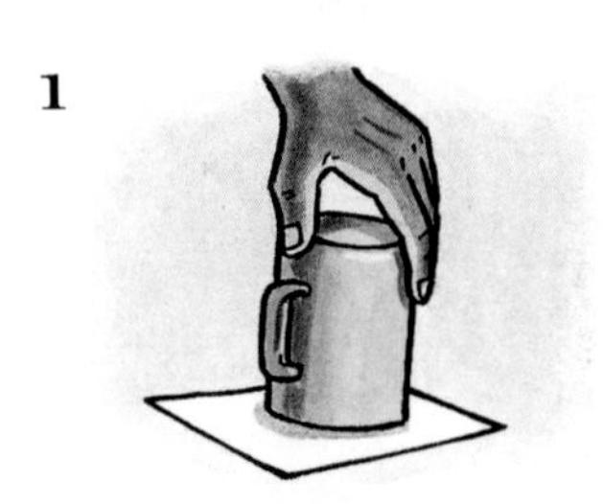

2

3

4

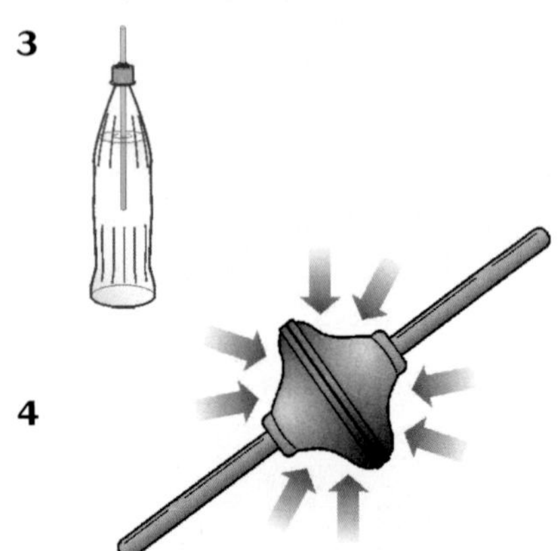

Figure 75.1 *Ways of demonstrating air pressure.*

Atmospheric pressure in action

These activities show you how strong atmospheric pressure is, and that it acts in all directions. In the first activity, the air pressure pushing on the paper is greater than the pressure of the water in the mug, and so it holds the paper in place.

For the second activity, before you heated up the water in the can, the air pressure inside the can was the same as the air pressure outside it. Boiling the water drove most of the air out of the can (you will see why in the next unit), causing a pressure difference. The air pressure was now greater outside the can than inside, and it pushed the sides of the can inwards.

In the third activity, you saw how you can't drink though a straw without atmospheric pressure. Normally, when you suck on a straw, you lower the pressure inside the straw. The air presses down on the water or drink in the glass, and because there is now a pressure difference (i.e. atmospheric pressure is higher than the air pressure inside the straw), it pushes the liquid up the straw. (A siphon works in the same way.) When you sealed the bottle, atmospheric pressure couldn't act on the water.

Science Extra

We don't feel the crush of atmospheric pressure because the pressure inside our bodies is almost the same as the pressure outside. But in space there is no air pressure and astronauts must wear pressurised space suits. If they didn't, they would explode.

1

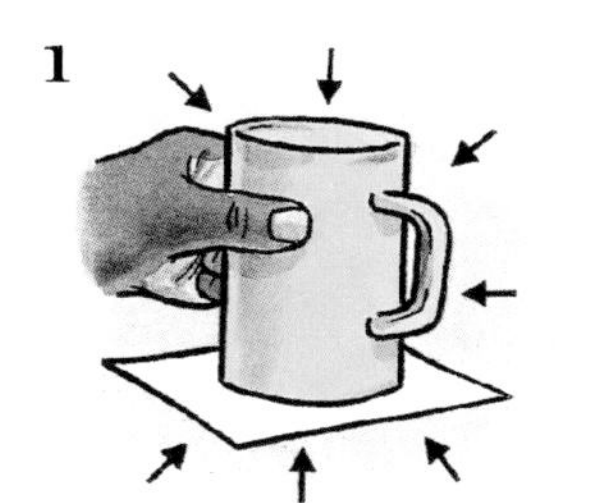

2

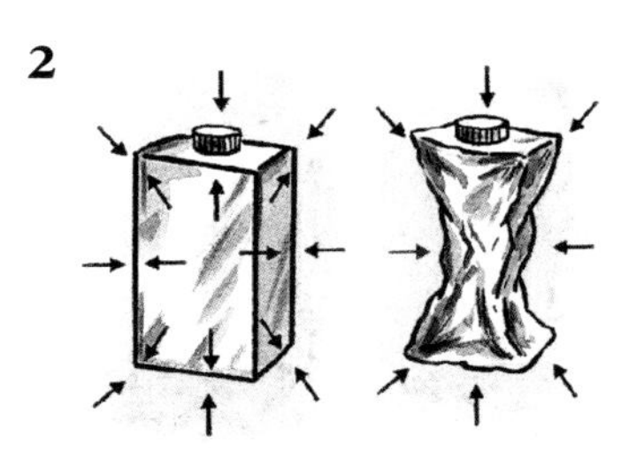

3

Figure 75.2 *Air pressure acts in all directions.*

Measuring air pressure with a barometer

Have you noticed the weather presenter on TV talk about high and low pressure systems? Weather forecasters use the changes in air pressure to predict the weather. A drop in atmospheric pressure usually means wind and rain. High atmospheric pressure usually means good weather.

We measure changes in air pressure using an instrument called a **barometer**. A mercury barometer consists of a long glass tube of mercury. (Can you think of the reason why mercury is used, rather than water?) As the air presses down on the surface of the dish, more mercury is forced up the tube. When the air pressure drops, less mercury can be supported, and the column of mercury gets shorter.

Unlike a mercury barometer, an **aneroid barometer** doesn't use liquid. *Aneroid* means *no liquid*. It consists of a sealed metal case, containing air at low pressure. If the atmospheric pressure rises, the top and bottom of the case are pressed in more.

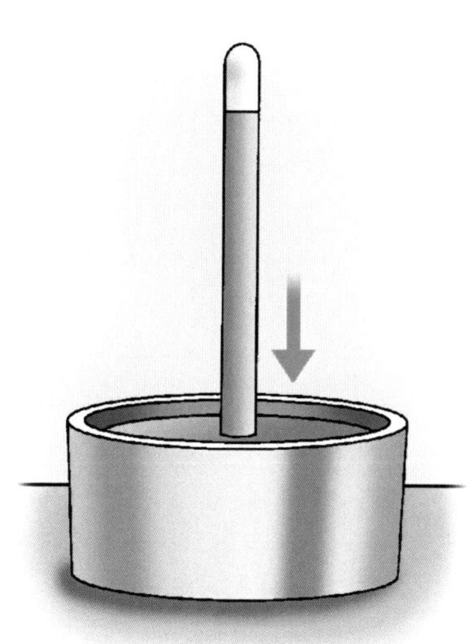

Figure 75.3 *A mercury barometer.*

Figure 75.4 *An aneroid barometer.*

Assessment

Measure the air pressure twice a day, or collect pressure (barometric) readings for tropical storms and hurricanes from the newspaper or local weather office. Compare these readings with the weather.

76

Pressure in gases

Keywords

compress, expand

Did you know?

Natural methane gas in Trinidad and Tobago is transported at high pressure from the gas fields to the National Gas Company (NGC) by underground pipelines. The mined gas is compressed about 600 times.

Cars move along on tyres pumped full of air. All gases have weight and so exert pressure. In the next activities you will see how you can increase the pressure of air or gas.

Figure 76.1 *Two bars of tyre pressure is equivalent to two atmospheric pressures.*

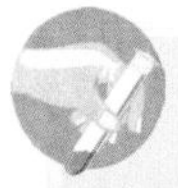

ACTIVITIES

1 Squashing air

- Put your finger over the end of a syringe that is full of air, making it airtight.
- How far you can push the plunger?
- What happens if you let go of the plunger?

2 Bursting a balloon

- Blow air into a balloon.
- Keep blowing, until it pops.
- Why does it pop?

3 Heating air in a balloon

- Stretch a balloon over the neck of a bottle.
- Place the bottle in a bowl of hot water and leave it for 5 minutes.
- What happens to the balloon?

1

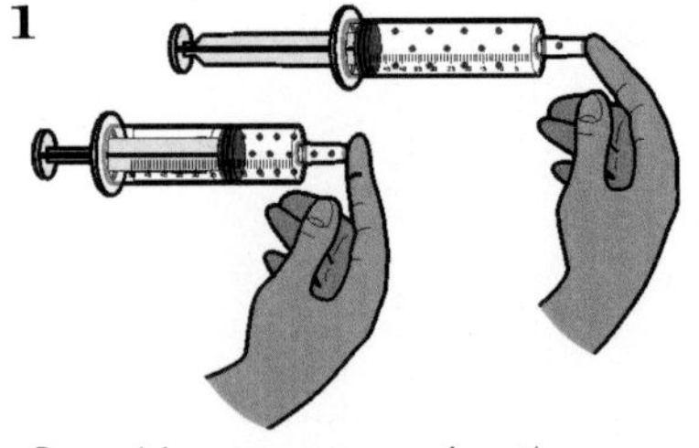

Squashing or compressing the gas.

2

Increasing the amount of gas.

3

Heating the gas.

Figure 76.2 *Ways of increasing the pressure of a gas.*

Ways of raising the pressure of a gas

Gases are easy to squash or **compress** because the molecules are far apart. You can increase the pressure of a gas by:

- *making the volume of the container smaller.* You did this by pushing in the plunger of the syringe. You also saw how the compressed air exerts a strong pressure – it pushed back the plunger when you let go.
- *putting more gas in the container.* You did this when you blew up the balloon. In fact, you filled the balloon with so much air that the pressure became too great, and it burst. But think of pumping air into a tyre. You can fit a lot of air in a tyre!
- *heating the gas.* You did this by heating the air inside the bottle. As the air heated, it **expanded**. The molecules of warm air moved faster and further apart, and this inflated the balloon.

Ways of using air pressure

Air pressure can be useful. Here are some examples.

- Air under pressure is a source of energy. A pneumatic drill is operated by compressed air. *Pneumatic* comes from an ancient Greek word meaning *air* or *breath*.
- Flight is made possible by differences in air pressure. The curved shape of the wings of an aircraft means that air passing over the top of the wings must travel further than the air under the wings. As the air must travel further, it must also travel faster, and this lowers the air pressure. The reduced air pressure causes an upward thrust under the wings, holding the plane in the air.
- Pressure cookers and autoclaves are designed to trap the steam produced by boiling water. This increases the pressure inside the pot or chamber, and raises the boiling point. Food cooks faster and bacteria are killed more easily for purposes of sterilisation.
- Hovercraft use compressed air to push downwards, so that the hovercraft moves just above the surface of the water.

Figure 76.4 *A pneumatic drill.*

Figure 76.3 *Hovercraft.*

Science Extra

A can of aerosol deodorant is a compressed gas in a liquid state. When you spray it, it feels cold. This is because the gas loses heat as it expands when it leaves the can.

ACTIVITY

Here's a simple device that makes use of the hovercraft principle.

- Glue a bottle top that has a pull-push valve to the centre of an unwanted CD.
- Push the valve of the top closed.
- Blow up a balloon and twist the end closed.
- Stretch the mouth of the balloon over the bottle top.
- Place the CD on a smooth, flat surface.
- Twist the balloon open and open the valve of the bottle top.
- Watch the CD skim along, powered by the release of air under pressure.

Figure 76.5 *Glue a bottle top with a pull-push valve in the centre of a CD.*

Assessment

Design any device that works using air pressure. Your teacher will judge the best design.

Pressure in liquids

Keywords
hydraulic

Did you know?
Over 80% of machinery on an oil rig is operated by hydraulic systems.

Like gases, liquids also exert pressure. Here are a few activities to try out.

ACTIVITIES

1 Pressure on all sides
- Use a nail to make holes, at the same height, around the circumference of a plastic bottle or bag.
- Quickly fill the bottle or bag with water.
- In what direction does the water come out of the holes? What does this tell you about the direction of the water pressure?

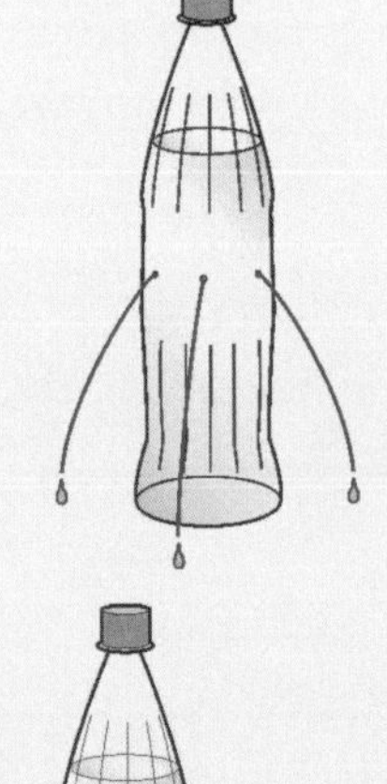

2 Pressure down below
- Make three holes on one side of a plastic bottle or bag – near the top, middle and bottom.
- Quickly fill the bottle or bag with water. Which hole does the water come out of the fastest? Why?

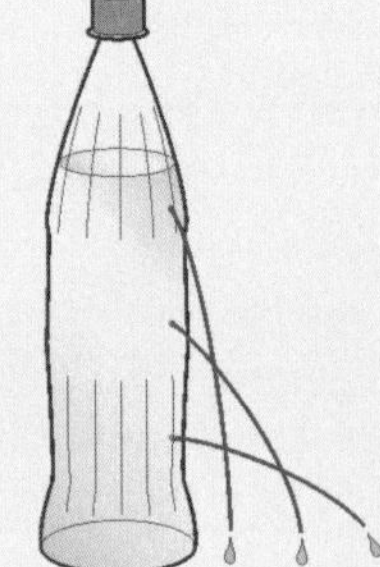

Pressure acts in all directions

In the first activity you saw how water came out of all the holes, because the water presses on all sides of the bottle. The pressure is perpendicular to the surface. That is why the water under pressure travels sideways a short distance, before it falls downwards because of the pull of gravity.

Figure 77.1 *Liquid pressure acts in all directions and increases with depth.*

Pressure increases with depth

In the second activity, you saw how the water spurted out the strongest, and furthest, from the bottom hole. The water at the bottom of the bottle has the most weight of water above it. Therefore the water pressure at the bottom is greatest.

The pressure in a liquid increases as depth increases. Have you ever dived down deep in a swimming pool or the sea and noticed the pressure on your eardrums? At only 10 m under water, the pressure is double that at the surface. For every 10 m deeper, the pressure increases by one atmosphere pressure. This explains why dam walls are built wider at the base.

Hydraulic systems – how liquids can transmit pressure

When a brake pedal is pushed down in a moving car, the car stops. Have you ever wondered how this works? Did you know that the car brakes work with the help of liquid pressure? The pressure applied with the foot is carried through the liquid in pipes to the wheels.

We call a system that uses liquid pressure, a **hydraulic** system. Hydraulic systems are used in machines such as jacks, lifts and forklift trucks. *Hydraulic* means *water pipe*.

This is how a car's brakes work: The brake pedal is pushed down with a small force over a long distance. At the brake this is turned into a large force that acts over a short distance to stop the car. In this way the force is amplified. This is how:

Remember that pressure = $\frac{\text{force}}{\text{area}}$.

If you rearrange this formula, force = area × pressure.

So if the area on which the pressure acts increases, the force increases too.

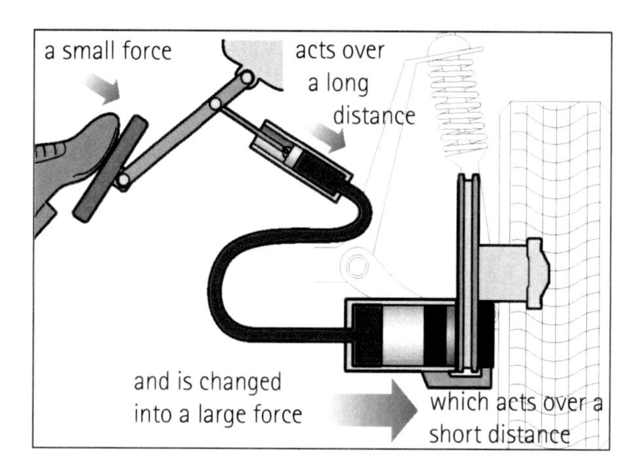

Figure 77.2 *How a car's brakes work.*

ACTIVITY

Hydraulics make use of the fact that liquids can't be compressed.

1 Half-fill a large and a small syringe with water.
2 Connect a tube to one of the syringes and press the plunger to fill the tube with water.
3 Connect the other end of the tube to the other syringe, making sure there is no air in the tube.
4 Push water from the large syringe to the small syringe. Gently hold your thumb on the plunger of the small syringe to feel the force. Measure the distance you push the plunger in the large syringe, and how far the plunger in the small syringe moves.
5 Then repeat, pushing water from the small syringe to the large syringe. Which system produces a stronger force?

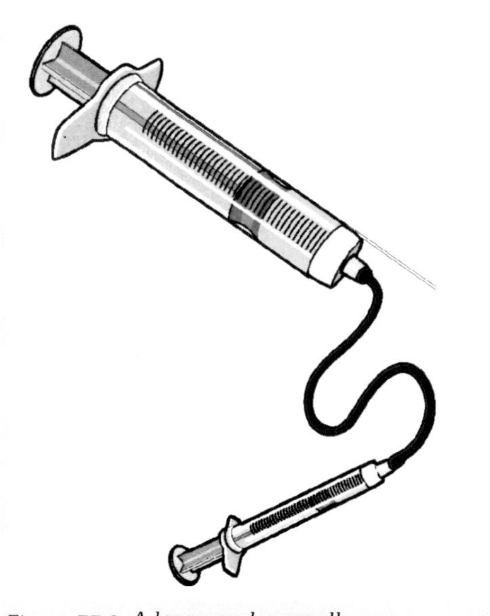

Figure 77.3 *A large and a small hydraulic system.*

Assessment

Place these sentences in the correct order to explain how pressure in a liquid is used in the braking system of a vehicle.

1 The liquid carries the pressure to the large piston.
2 The large piston carries a big force to the tyres.
3 The small piston is pushed down.
4 The large piston pushes up with the same pressure as the small piston was pushed down with.
5 The small piston applies pressure to the liquid.

Science Extra

Some bacteria live in high-pressure environments, and are called *barophiles*, meaning *pressure-loving*. The ***barophiles*** that live on the deep-sea ocean floor can survive pressures that are 380 times atmospheric pressure or more.

78

Review topic 5

There are many ways we can define and describe forces. A force is a push or a pull. It has size and direction, and can therefore be measured (in newtons). A force is an interaction between two objects. It can act between objects in contact (a contact force such as friction), or between objects at a distance (a non-contact force such as gravity).

With a proper understanding of what a force is, we can understand the difference between weight and mass. Earth pulls on objects with a gravitational force that is equal to the weight of an object. Forces have effects: they can transfer energy, and they can change the shape of objects – permanently if the materials are non-elastic, and temporarily if the materials are elastic. More importantly, a force can make an object move, or stop its movement. A force can speed up (accelerate) an object, slow it down, or make it change direction. Friction is an example of a force that can act in the opposite direction to the movement of an object.

Inertia and momentum are properties that relate to the movement of an object, and both depend on the mass of the object. Inertia is the reluctance of an object to start moving or stop moving. Momentum is a measurement of the movement of an object – the product of its mass and velocity.

Pressure is force acting on area. The unit of pressure is newtons per metre squared, or the pascal and it's a measure of how concentrated a force is. Solids (e.g. wooden blocks), gases, and liquids all exert pressure. Atmospheric pressure is an example of pressure from the layer of air around us. Pressure in liquids can be transmitted, using the principle of hydraulics.

Multiple-choice questions

1 Which of the following statements is incorrect? A force:

a can act from a distance.

b has an equal and opposite reaction force.

c has size and direction.

d is needed to keep an object moving at a steady speed in a straight line.

2 The type of force that resists motion is called:

a gravity

b momentum

c friction

d inertia

3 Which of the following statements about pressure is correct?

a pressure = force × area

b Pressure is higher at the top of a column of water than at the bottom.

c Pressure can be transmitted through liquids.

d The unit of pressure is Pa.

Longer questions

1 The table below shows results from an experiment investigating the stretching of a rubber band with various weights.

Weight (N)	Extension (m)
0	0
10	2
20	4
40	8
65	14
80	16
100	20
110	22

a Draw a graph of extension against load using the data from the table.

b What conclusion can be drawn from this experiment?

c What is the name of the law that describes this observation?

2 Draw a wheelbarrow being pushed and show, with labelled arrows, the three main forces acting on it.

3 a An astronaut has a mass of 65 kg.

i What would her weight be on Earth?

ii What would her weight be on the Moon?

b Is gravity the same value everywhere on Earth?

4 a What is friction?

b Give three advantages of friction.

c Give three disadvantages of friction and ways in which it can be reduced.

5 Explain why you continue to move forwards in a vehicle when the brakes are suddenly applied. What safety feature counteracts this property?

6 A truck, a car and a bicycle are all moving with a speed of 20 m/s. Which vehicle possesses the greatest momentum? Explain your answer.

7 Two cars are travelling at 30 m/s. One has a mass of 800 kg, the other a mass of 1 100 kg. Which one will need a bigger force to accelerate it to 35 m/s? Explain your answer.

8 A man has a weight of 920 N. His feet cover an area of 0.3 m^2. A woman has a weight of 620 N and her feet cover an area of 0.2 m^2. Which of them exerts the greater pressure on the floor?

9 Explain how and why pressure varies on a person

a on top of a mountain,

b at sea level, and

c scuba diving 20 m under the sea.

10 A 2 litre plastic bottle is filled with water and four holes are made (one at 8 cm from the base, two at 4 cm from the base and one at 2 cm from the base of the bottle).

a Draw a diagram to show how water will squirt out of the bottle.

b Account for the different distances the water will travel from each hole.

11 Draw a design of a hydraulic arm that uses syringes to lift a box.

Properties of acids and bases

Keywords

acids, alkali, bases, corrosive, neutral

Did you know?

Calcium dissolves in acid. If you soak substances that are strengthened by calcium (for example, an eggshell, a tooth, or a chicken bone) in vinegar for a day or two, they will dissolve.

What are acids and bases?

You already know that matter can be classified in different ways. In this unit you will find out how to group substances as **acids**, **bases** or **neutral substances**. Almost every substance is an acid or a base. One exception is distilled water, which is neutral. Distilled water is pure. It is without any solutes and contains no ions (charged particles).

Acids and bases are liquids that can be found everywhere. For example, acids and bases can be found in kitchens, medicine cabinets, garages and school laboratories. Even your own stomach contains an acid that helps to break down the food you eat before it passes into your intestines.

Figure 79.1 *Natural and man-made acids.*

Figure 79.2 *Natural and man-made bases.*

Warning!

The incorrect classification of household chemicals, foodstuffs and medicines can cause illness or death. The only acids you should ever taste are those found in foods. Never taste a chemical in the school science room, unless your teacher tells you it is safe to taste. Laboratory acids and bases are extremely dangerous. They can burn your skin and make holes in clothing. Use them carefully.

What are the properties of acids and bases?

Acids and bases have certain properties or characteristics that help us to identify them. Acids have a sharp, sour taste. Some are **corrosive**, which means they burn your skin or 'eat away' materials. Car batteries contain sulphuric acid, which is very corrosive. Acetic acid is the weak acid found in vinegar. Lemons, oranges and limes contain citric acid. Tea contains tannic acid. Sour milk contains lactic acid.

Bases feel soapy when they are dissolved in water. A base that dissolves in water is called an **alkali**. Soap contains an alkali that breaks down grease. This is why soap helps to clean things. Bases can be as corrosive as acids. Some household cleaners (oven cleaner, for example) contain concentrated amounts of bases and should be used with care. Bicarbonate of soda is a harmless base used in cooking and in some medicines. Bases are found in most cleaning products.

Some of the properties of acids and bases are shown in the table below.

Properties of acids	Properties of bases
can feel rough between the fingers neutralise a base change colour with various indicators	soapy and slippery feel alkalis neutralise acids change colour with various indicators

Science Extra

When you peel an onion, your eyes may start to water. This is because sulphur molecules drift through the air, from the cut onion into your eyes. A chemical reaction takes place between your salty tears and the onion juice. As a result, sulphuric acid forms.

INVESTIGATION Classify household chemicals

Work with a partner. You will need: a beaker of water, three saucers, dishwashing liquid, orange juice and vinegar.

- Number and label each saucer 1–3.
- Mix a little of each substance with 5 ml of water in a saucer.
- Feel all the the substances using your thumb and middle finger.

1 Copy the table below. Fill in the table with each substance you test, once you have decided how to classify it. Record your observations and results as well. The properties table above will help you to classify each substance.

Acid	Base	Alkali (base that dissolves in water)

2 Compare your findings with another pair of students. Talk about the following: Which substances felt soapy or slippery?

Assessment

- Which acids are contained in fruit and vegetables? List the citrus fruits that you might find in the Caribbean. Make a poster identifying them.
- Research the following different acids: citric acid, lactic acid in sour milk, ethanoic acid, tartaric acid, ascorbic acid, formic acid and oxalic acid. Report your findings to the class.

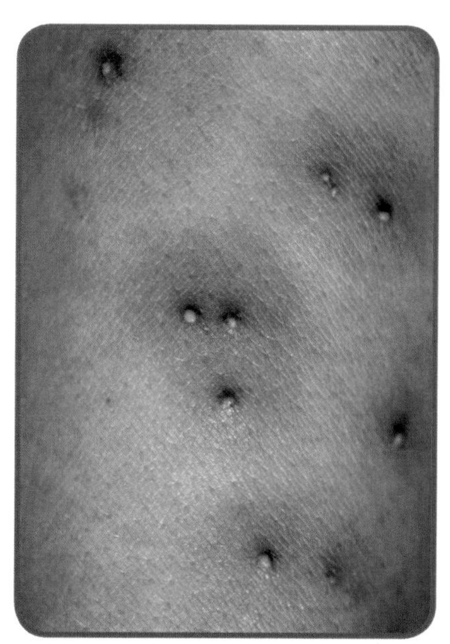

Figure 79.3 *An ant bite contains formic acid, which causes pain and swelling. This is what fire ant rash looks like 24 hours after stinging.*

80

Indicating colour changes

Keywords
acidic, acidity, alkalinity, basic, indicator, litmus paper

Did you know?

An easy way to test whether a substance is an acid or an alkali is to use **litmus paper**. Litmus paper is absorbent paper that has been soaked in a dye solution taken from lichen (a fungus). The paper is then dried. Litmus paper reacts in the following ways:

- It turns red when dipped into an acid.
- It turns blue when dipped into an alkali (a base).
- It does not change colour when dipped into a neutral substance.

Effects of acids and bases on indicators

Many acids and alkalis are too dangerous to taste, or even to touch. A safer and more accurate way of testing them is to use an **indicator**. Indicators are dyes that change colour when acids or bases are added to them. If we want to find out whether a substance is **acidic** or **basic**, we can use the following:

- a natural indicator like tea or red cabbage water
- a chemical indicator like bromothymol blue (BTB) or methyl orange.

INVESTIGATION Use a chemical indicator

Work in groups. You will need: basic substances, acidic substances, a neutral substance, water, bromothymol blue (BTB), test tubes or glass jars.

- Look at Figures 79.1 and 79.2 in Unit 79. You are going to test some of these common substances. Before you begin, predict whether the substance is an acid, base or neutral substance.
- Number and label the glass jars. Pour 5 ml of water into each jar. Add five drops of BTB to each. BTB should have a greenish colour if the water is pure.
- Taking one glass jar at a time, add a small amount of acid or base solution until you have tested them all.

1 Record the colour changes that you see. You can copy and fill in a table like the one below.

Jar or test tube number	Substance	Colour change of BTB
1	salt	the colour changed from green to ...
2	vinegar	

2 Which substances change the BTB indicator to yellow? Which substances change it to blue? Which substances don't change the colour of the BTB indicator? Which of the liquids were acids? Which of the liquids were bases? Which were neither acidic nor basic?

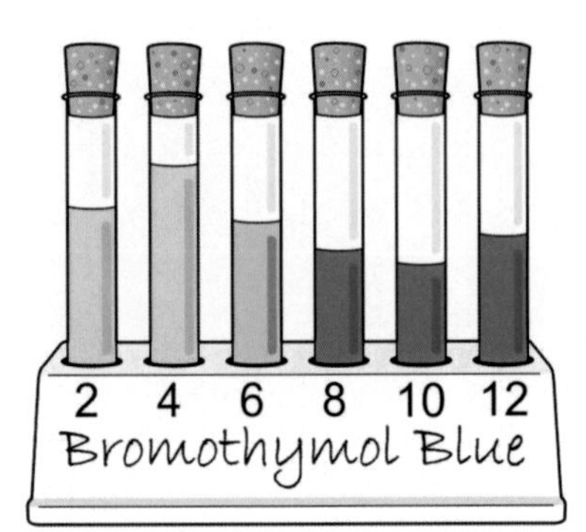

Figure 80.1 *Bromothymol blue is a commercial chemical indicator. It is yellow in acidic solutions, green in neutral solutions and blue in basic solutions.*

Figure 80.2 *Lichen is used in making litmus paper.*

INVESTIGATION Make a natural indicator

Work with a partner. You are going to make indicators from common plants such as hibiscus petals, sorrel and red cabbage. Then you can test a range of common substances for **acidity** or **alkalinity**. You will need: 3 bowls, sieve, scissors, chopping board, three glass jars, saucer, knife, measuring jug, boiling water, notebook, pen, sticky labels.

Figure 80.3 *sorrel* *red cabbage* *hibiscus flower*

- Finely chop or crush some hibiscus petals, sorrel and red cabbage leaves.
- Put the chopped plants in separate bowls. Pour boiling water over them. Leave to soak until the water changes colour. Remember to be careful when you are working with sharp instruments or boiling liquids.
- Let the mixtures cool down. Hold the sieve over the jug. Pour the cabbage water through the sieve. Now pour it from the jug into a glass jar, which you must label 'red cabbage'. Repeat this process with the liquids (filtrates) from the other two bowls. You should have three jars of liquid, labelled 'hibiscus petals', 'sorrel' and 'red cabbage'. These are your natural indicators.
- Your teacher will give you test tubes of acid and alkali. Test what colour your three indicators turn in acids and in alkalis by adding a few drops to the test tubes. Copy and complete the following table.

Indicator	Colour in acid	Colour in alkali
hibiscus petals		
sorrel		
red cabbage		

Science Extra

Many years ago, people used organic or natural dyes like crushed leaves and berries. The crushed matter was then rubbed on to surfaces such as leather, paper or pottery. It was also used for face and body decoration.

Later, people began to dye cloth. Dyes can be classified as direct dyes, vat dyes, acid dyes or dispersed dyes. Acid dyes are used for colouring protein fibres, such as wool and silk. These dyes were originally applied in an acidic bath. Mordants are chemicals that bind (fix) a dye to the fibres in a material. Examples of common mordants include cream of tartar (potassium hydrogen tartrate), salt (sodium chloride), and baking soda (sodium bicarbonate).

Figure 80.4 *Some materials are made using batik dyes.*

Assessment

- Write a set of instructions on how to make a natural indicator. Compare your own instructions with another student's work. Are the instructions similar?
- Research the acid dyeing process. Find out about fabrics used in carnival costumes. Look at the two websites: http://www.dyeman.com/natural%20Dye%20recipe.htm and http://www.joaniemitchell.com/batik-painting/process.html
- Look through back copies of *Caribbean Beat* magazine. Collect some pictures of the various Trinidadian carnival costumes including those of Mas, Moko Jumbies and Minstrels. Find out about the costume designer Peter Minshall.

81

Using the pH scale

Keywords

litmus, pH scale, universal indicator

Did you know?

The acid in your stomach is hydrochloric acid. It has a pH of about 1.4, making it strongly acidic. The lining of the stomach is protected against corrosion by a thick layer of mucus.

pH	Substance
14	liquid drain cleaner, caustic soda
13	bleaches, oven cleaner
12	soapy water
11	household ammonia (11.9)
10	milk of magnesia (10.5)
9	toothpaste (9.0)
8	baking soda (8.4), seawater, eggs
7	pure water (7)
6	urine (6), milk (6.6)
5	acid rain (5.6), black coffee (5)
4	tomato juice (4.1)
3	grapefruit, orange juice, soft drink
2	lemon juice (2.3), vinegar (2.9)
1	hydrochloric acid secreted from the stomach lining (1)

Figure 81.2 *The pH scale and corresponding colours of universal indicator.*

Indicators and the pH scale

An indicator is a substance that turns a certain colour depending on the pH of a solution. **Universal indicator** is a mixture of indicators. It has different colours across the whole pH range and it is used to indicate the pH of many substances. It tells us exactly how strong or weak an acid is. You can use **litmus** and universal indicator in a paper form.

The table below shows the colour changes that take place in different indicators when they are placed in acidic or alkaline solutions.

Figure 81.1 *The litmus paper test.*

Indicator	blue litmus paper	red litmus paper	methyl orange	phenolphthalein
In acid	red	no change	red	colourless
In alkali	no change	blue	yellow	pink

The pH scale measures acidity and alkalinity

Scientists invented a sequence of numbers called the **pH scale** to measure the acidity or alkalinity of a substance. The strengths of acids and alkalis are represented by the letters p and H, followed by a number between 1 and 14. The number 1 shows strongest acidity and 14 shows strongest alkalinity. In the middle of the scale, 7 indicates neutral substances such as water. We say that orange juice, for example, has a pH of 3.

ACTIVITY

Use Figure 81.2 to answer the questions. Write down your answers.

1 Is a substance with a pH of 6 acidic or alkaline?
2 Is a substance with a pH of 3 a stronger or a weaker acid than a substance with a pH of 5?
3 Is a substance with a pH of 7 acidic or alkaline, or neither?
4 Which do you think is more dangerous: an acid with a pH of 5, or an alkali with pH of 13? Give reasons for your answer.
5 Record the pH of the following substances: acid in the human stomach, acid rain, saliva, milk, fizzy drink, seawater, ammonia, dishwashing liquid.
6 Which is the stronger acid, lemon juice (pH of about 2.3), or black coffee (pH of about 5.0)?
7 The pH of caustic soda is 14 and the pH of toothpaste is about 10. Which is the stronger base (alkali)?

INVESTIGATION

Find the pH values of common substances

For this investigation you will need universal indicator paper and solutions of the following substances: hydrochloric acid, vinegar, toothpaste, fizzy drink, lime juice, tap water, milk, toilet bowl cleaner, curry powder, bleach, dishwashing liquid, black tea.

- Place a few drops of each solution on a strip of universal indicator paper. Make a note of the colour change of the indicator.
- Record your findings in a table like the one below. Classify each substance as strong acid, weak acid, neutral (or almost neutral), weak alkali or strong alkali.

Substance	Colour of universal indicator	pH value	Classification
lime juice	pale orange	4	weakly acidic

Figure 81.3 *Some acids are stronger than others. Hydrofluoric acid can be used to dissolve glass to create a picture on its surface.*

Assessment

Write a short article for the magazine *Caribbean Farmer's Weekly*. Report on environmental damage to local crops such as tobacco, sugarcane or tea caused by emissions from the nearby fertiliser and petroleum industries. Include the following in your article: information on pH testing of the soil, possible solutions to the problem and research on how to improve soil quality.

Neutralising substances

Keywords

ion, neutralisation, product, reactant, salt, sodium chloride, water

Did you know?

Industry uses neutralisation reactions to manufacture salts and fertilisers. Neutralisation reactions are useful for health purposes too. For example, we use antacid tablets to neutralise excess stomach acid that causes heartburn, and we use toothpaste to neutralise the acids in our mouths that cause tooth decay.

Neutralisation

Your investigations in Topic 6 have shown that, in some ways, acids and bases are opposites. What do you think would happen if you mixed an acid and a base together? When an acid and a base react together, both solutions lose their properties. They 'cancel each other out'. The scientific word for this is **neutralisation**.

When an acid and a base react, the H^+ **ion** from the acid and the OH^- ion from the base react to create **water** (H_2O). The remaining ions of the acid and the base combine to form a neutral **salt**. The reaction can be written as a word equation:

acid + base → salt + water

For example, when hydrochloric acid reacts with sodium hydroxide in a neutralisation reaction, the hydrogen ion and the hydroxide ion combine to make water. The chloride ion and the sodium ion stay as they are. These remaining ions combine to create **sodium chloride**, better known as table salt. The chemical word equation for this reaction is:

hydrochloric acid + sodium hydroxide → sodium chloride + water

The same equation can be written using chemical formulae (symbols) instead of words:

$HCl + NaOH \rightarrow NaCl + H_2O$

INVESTIGATION Do a taste test

You will need: lemon juice, bicarbonate of soda, clean medicine dropper or teaspoon.

- Put two drops of lemon juice on your tongue. How does it taste?
- Put two more drops of lemon juice on your tongue followed by a pinch of bicarbonate of soda.

1 How has the taste of the lemon juice changed?
2 Why do you think the sour, acidic taste altered?

ACTIVITY

We can also carry out neutralisation reactions in the laboratory. Remember, when acids and alkalis react together they form a chemical called a 'salt' plus water.

- On the next page is the range of common acids and alkalis found in the lab. Can you match the **reactants** (in the left-hand column) to the

products (in the right-hand column)? One has already been done for you. Check with your teacher and then copy the correct neutralisation reaction into your books.

Reactants		Products
hydrochloric acid (HCl) + potassium hydroxide (KOH)		sodium carbonate (Na_2CO_3) + water (H_2O)
sulphuric acid (H_2SO_4) + ammonium hydroxide (NH_4OH)		potassium chloride (KCl) + water (H_2O)
nitric acid (HNO_3) + calcium hydroxide ($Ca(OH)_2$)		calcium nitrate ($Ca(NO_3)_2$) + water (H_2O)
carbonic acid (H_2CO_3) + sodium hydroxide (NaOH)		ammonium sulphate ($(NH_4)_2SO_4$) + water (H_2O)

- Write word equations for the following reactions:
- Nitric acid reacts with magnesium hydroxide.
- Sulphuric acid reacts with sodium hydroxide.
- Hydrochloric acid reacts with lithium hydroxide.

INVESTIGATION Make neutral solutions

Work with a partner. You will need: two glass jars, an acid, an alkali, glass rod to stir, litmus paper, tartaric acid, water, bicarbonate of soda, a natural indicator. The steps in this investigation are set out in a flowchart.

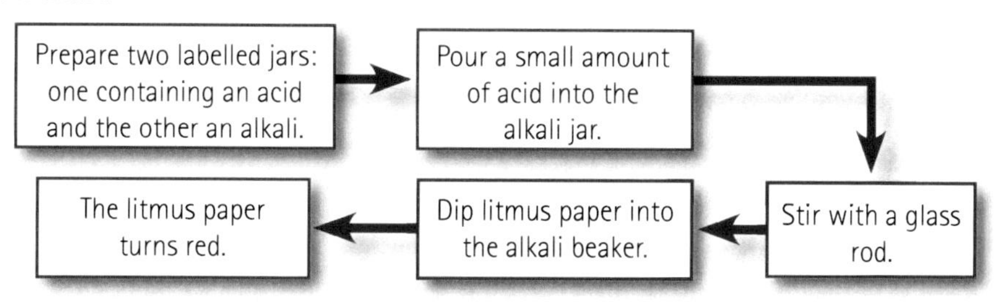

1 Explain the steps in the investigation to your partner and describe what happened.
2 Do the experiment again using half a teaspoon of tartaric acid mixed with half a beaker of water in the first jar and half a teaspoon of bicarbonate of soda in the second jar. Use a natural indicator (hibiscus petals, sorrel or red cabbage) instead of litmus paper.

Assessment

You can refer to the pH scale in Unit 81.

- Research how plants and animals have evolved to produce acids and alkalis to protect themselves.
- Design, draw and label an imaginary animal or plant that uses either an alkali or an acid as a survival mechanism. Show your design to the class. Talk about ways of neutralising the effects of the mechanism.

Bee

Fire coral

Purple water grass

Macuna sea beans

Figure 83.1 *Do you recognise these animals and plants? They all produce acids or alkalis to protect themselves.*

83

Reacting acids

Keywords

carbonates, carbon dioxide, chemical reactions, compounds, hydrogen gas

Did you know?

Strong acids release a large number of H^+ ions. A weak acid releases fewer H^+ ions.
When a strong acid (sulphuric acid, for example) is added to water, a lot of heat is released. The heat given off can even break the bottle. So remember, it is safe to add acid to water, but NEVER add water to acid. Always wear safety goggles when you work with acid to stop it getting into your eyes.

Reactions of acids

In this unit you will learn about some of the ways in which acids react with other **compounds**. A car going rusty in the rain, a cake baking in the oven, fireworks exploding at carnival time – these are all examples of **chemical reactions**.

Figure 83.1 *A chemical reaction causes fireworks to explode in the air.*

Chemical reactions take place when one substance combines with another. As a result, a new substance is created. Chemical reactions happen around us, and even inside us. Think about the many reactions taking place in our bodies that keep us alive.

The table below shows some common laboratory acids, along with their chemical formulae and what they can be used for. All acids contain hydrogen atoms (H) bonded to other non-metal atoms.

Acid	Formulae	Use
hydrochloric acid	HCl	cleans cement off bricks, controls the acidity of swimming pools
sulphuric acid	H_2SO_4	used in motor car batteries
nitric acid	HNO_3	used to make fertilisers and explosives

Acids react with most metals. The reaction produces a salt and **hydrogen gas**. In a laboratory investigation, the hydrogen gas should make a 'popping' noise when a lighted match is held at the opening of the container. When acids react with metal or **carbonates**, salts are formed. You will learn more about salts in Unit 84.

INVESTIGATION Find out how acids react with metals

Work in a group. You will need: three test tubes, small piece of zinc, small iron nail, dilute sulphuric acid or hydrochloric acid, matches, candle, Bunsen or spirit burner.

- Fill two test tubes one-third full with either dilute sulphuric acid or dilute hydrochloric acid. Clean a small iron nail and a piece of zinc with steel wool and put them in the acid.

1. Which metal reacts faster? Does the acid change colour?
   - Touch the test tube. Is it warm? If a reaction seems slow, heat the acid gently over a lit candle, a Bunsen or a spirit burner.
   - Collect the gas that is formed by inverting an empty test tube over the reaction test tube. Hold a lighted match at the mouth of the test tube.
2. What happens? Record your conclusion.
3. Working in pairs, try to complete a chemical word equation for the reactions below:
   zinc and sulphuric acid
   iron and hydrochloric acid

Remember, the general reaction is: metal + acid → salt + hydrogen
And the word equations are:
zinc + hydrochloric acid → zinc chloride + hydrogen
iron + sulphuric acid → iron sulphate + hydrogen

Science Extra
A chemical extinguisher can be made by combining an acid and a carbonate. The **carbon dioxide** produced will extinguish a fire.

INVESTIGATION Find out how acids react with metal carbonates

Work in a group. You will need: a large test tube, calcium carbonate (powdered chalk) or sodium carbonate or sodium hydrogencarbonate (bicarbonate of soda), dilute hydrochloric acid or sulphuric acid or vinegar, clear limewater in a test tube.

- Fill a quarter of a test tube with a carbonate. Add a few drops of acid. Record what you find out.
- $CaCO_3$ is the formula for calcium carbonate. Which gas would you expect to get from it? To give you a clue, it is a gas that you breathe out.
- Add more acid to the carbonate and test for the gas. What is your conclusion? This is also a neutralisation reaction. Write the chemical equation for it: carbonate + acid → salt + carbon dioxide + water

Figure 83.2 *A lot of the limestone on a coral reef has been produced over millions of years by the tiny life forms that produce coral (calcium carbonate).*

Assessment

- Read pages 12 and 13 again.
- Think about, and then draw, a marine food chain or web.
- Most pollutants that enter aquatic environments come from human activities. For example, waste from factories or refineries, sewage from treatment plants, detergents from washing, and pesticides and fertilisers used in agriculture. Carry out research into how Caribbean coral reefs are damaged by industrialisation and tourist liners. What impact will the loss of coral reefs have on marine life and the economy of the islands?

84

Making salts – solutions and solubility

Keywords

salts, solubility, solute, solution, solvent

Did you know?

If you crave salt, you may lack certain minerals or sodium chloride in your body.

Salts

You have already learnt that a salt is formed when a metal replaces the hydrogen of an acid. **Salts** are substances that form when acids react with metals or alkalis. Copper sulphate and calcium carbonate are examples of salts.

Do you remember from Unit 82 that when an acid and a base react, the H^+ ion from the acid and the OH^- ion from the base react to create water, and that the remaining ions combine to form a neutral salt? The reaction can be written as an equation. The last part of the salt's name depends on the acid from which it was made. The full name of a salt depends on the acid and the substance with which it reacts. You already know that:

acid + base → salt + water.

Let's look at some other examples.

- When acid reacts with metals:
  acid + metal → salt + hydrogen gas
  hydrochloric acid + zinc → zinc chloride + hydrogen gas
- When acid reacts with bases or alkalis (the neutralisation reaction):
  acid + base/alkali → salt + water
  hydrochloric acid + calcium hydroxide → calcium chloride + water
- When acid reacts with carbonates:
  acid + carbonate → salt + water + carbon dioxide
  hydrochloric acid + calcium carbonate
  → calcium chloride + water + carbon dioxide

Figure 84.1 *All seas are salty because sodium chloride is dissolved in the water. The Dead Sea in Israel is famous because the water is so salty that people can float very easily in it.*

The table below shows some common household substances that are all types of salt.

Common name	Chemical name of active ingredient	Type of chemical	Uses
baking powder	sodium hydrogencarbonate	salt	raising agent in baking treating insect bites
Epsom salts	magnesium sulphate	salt	laxative
toothpaste	sodium monofluorophosphate	salt	prevents tooth decay

The chemical name for ordinary, common salt is sodium chloride. Salt has many uses, such as flavouring and preserving food.

Figure 84.2 *A salt pan in Bonaire.*

Figure 84.3 *Table salt.*

Figure 84.4 *Coarse salt used to preserve food.*

Figure 84.5 *Fish that has been preserved in salt being hung up to dry.*

Solutions, solvents, solubility

There are three very similar words that scientists use when they talk about dissolving: solute, solvent, and solution. It is important to learn the different meanings of the words:

- **Solute** – a substance that dissolves in a liquid
- **Solution** – a liquid containing a dissolved substance
- **Solvent** – a substance in which other substances are dissolved, often a liquid.

The table below shows types of solutions. You will see that gases and liquids can be solvents too.

Type of solution	Solute	Solvent	Examples
solid in liquid	solid	liquid	sugar in water
solid in solid	solid	solid	alloys (mixture of metals) such as brass = copper + zinc
gas in liquid	gas	liquid	carbon dioxide in water (fizzy drinks)
liquid in liquid	liquid	liquid	alcohol or syrup in water
gas in gas	gas	gas	air = oxygen, carbon dioxide, nitrogen and noble gases

INVESTIGATION Test different solutes and solvents

Not all solutes dissolve in all solvents.

- Pour a small amount of each of the four solvents into separate test tubes.
- Add a small amount of sodium chloride to each solvent. Shake each test tube well and note what happens.

1 Record your results in a table like the one below. Repeat with fresh solvents for each of the other solutes in turn. Say which, if any, is the best solvent for each of the solutes.

Solute	Solvent			
	water	vegetable oil	kerosene	alcohol
powdered chalk				
sodium chloride				
butter				
cheese				
pitch (or bitumen)				
copper sulphate				

Solubility and temperature

How easily a solute can dissolve in a solvent is called its **solubility**. Salt has a high solubility in water. Butter has a high solubility in oil and a low solubility in water. The solubility of sodium chloride in water varies according to temperature.

INVESTIGATION What happens when a solvent is heated?

Do you think a solute will dissolve better in a hotter or a cooler solvent? Let's see if you are right.

You will need: a small container of water that has been kept at room temperature. Add one teaspoon of sodium chloride and wait until the salt dissolves.

INVESTIGATION What happens when a solvent is heated? (continued)

- Continue to add sodium chloride, one teaspoon at a time, until no more of the solute dissolves.

1 How many teaspoons of salt dissolved?
  - Repeat the first step, but with water heated to 35 °C, 45 °C, 60 °C, 70 °C and 80 °C.
  - Take care when using hot water.

2 Write down your observations. Record your results in a table.

3 Plot a line graph of your results. Show the number of teaspoons of sodium chloride dissolved (along the y-axis) against the temperature of the water (along the x-axis).

4 What conclusion can you make?

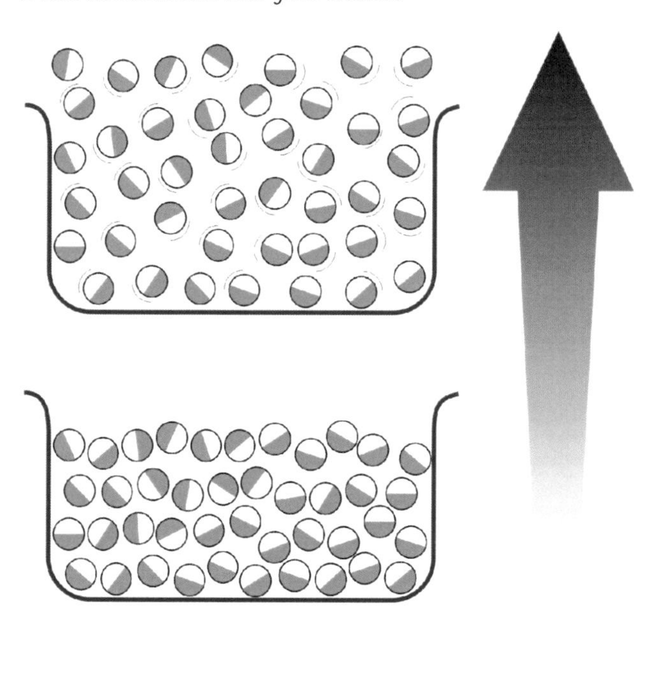

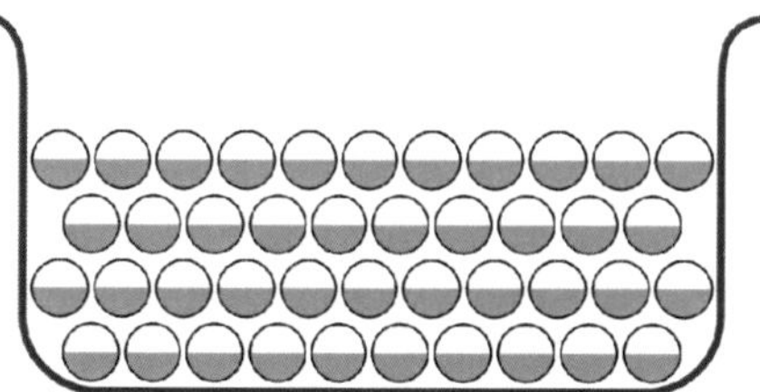

Figure 84.6 *When you heat a solvent, the particles in the solvent gain more energy and move faster and further away from one another.*

Assessment

Many of the Caribbean Islands are made up almost entirely of limestone. The limestone has built up over millions of years. It is formed from the skeletons of tiny organisms (sea creatures) that produce calcium carbonate (coral).

- Research acid rain, which damages limestone buildings.
- Imagine that you are an eco-activist. Prepare a campaign for the reduction of air pollution. As part of your campaign you must promote the use of alternative energy sources.

84

Review topic 6

Substances can be classified into acids and bases. Pure distilled water is neither an acid nor a base – it is neutral. Acids in foods produce a sharp or sour taste, while bases produce a bitter or salty taste. A base that can dissolve in water feels slippery and soapy, and is known as an alkali. Bases are used in most cleaning products.

Strong acids are found in our stomachs and in batteries. A strong base is bleach. Both strong acids and strong bases can corrode (eat away) materials or burn skin, so they must be handled with great care.

The strength of acids and bases is measured on a 14-point pH scale, with a pH of 7 being neutral. Substances with a pH below 7 are acids, and substances with a pH above 7 are bases. The further the pH drops below 7, the stronger the acid, and the higher the pH rises above 7, the stronger the base. Indicators can identify an acid or base by changing colour.

When acids react with substances they release H^+ ions. The stronger the acid, the more H^+ ions they release. Acids react with water and give off heat. Acids react with most metals and carbonates (carbon-containing salts) to produce salts. When an acid and a base come together they react to neutralise each other, producing water and a neutral salt.

Salts are named after the two substances they are made from – the acid and the substance it reacts with. Salts dissolve easily in water to form solutions. In a solution, the dissolved salt is called the solute, while the water is known as the solvent.

Multiple-choice questions

1 Which of the following is not an acid?
  a tea
  b vinegar
  c lime juice
  d baking powder

2 Sodium hydroxide is:
  a an acid
  b an alkali
  c a neutral compound
  d a gas

3 Which of the following is not an indicator?
  a methyl orange
  b water
  c red litmus
  d universal indicator

4 Which of the following groups can all be used to make natural indicators?
  a water, red cabbage, sorrel
  b red cabbage, sorrel, hibiscus petals
  c water, red cabbage, sorrel
  d water, sorrel, hibiscus petals

5 The pH scale ranges from
  a 1 to 14
  b 1 to 6
  c 7 to 14
  d 6 to 7

6 Which of the following pH values represents the strongest acid?
  a pH 2
  b pH 7
  c pH 1
  d pH 14

7 When a glowing splint or match relights in a sample of gas, the gas is most likely to be:
  a carbon dioxide
  b oxygen
  c nitrogen
  d water vapour

8 When a lighted splint or match causes an explosion in a sample of gas, the gas is most likely to be:
  a carbon dioxide
  b oxygen
  c nitrogen
  d hydrogen

9 Acid added to a metal will produce:
  a a salt and water
  b a salt and oxygen
  c a salt and hydrogen
  d water and carbon

10 Water is produced by reacting:
  a an acid and an alkali
  b an acid and a metal
  c a metal and an alkali
  d water and an alkali

11 Which of the following chemical formulae represents a salt?
  a NaOH
  b NaCl
  c HCl
  d KOH

12 When a solute dissolves in a solvent it produces a:
  a stronger solute
  b solution
  c stronger solvent
  d weaker solvent

13 To remove tar from a shovel, we can use:
  a water
  b sodium chloride
  c kerosene
  d soap

14 The solubility of a solute in a solvent depends on:
  a the amount of water vapour in the air
  b the temperature of the solvent
  c the size of the container
  d the time taken to mix the solution

Longer questions

1 List five properties of acids and bases in a table like the one below.

Properties of acids	Properties of bases

2 What is an acid?
  a Give three examples of acids.
  b List three uses of acids.

3 What is an alkali?
  a Give three examples of alkalis.
  b List three uses of alkalis.

4 You are given three colourless liquids labelled A, B and C. You are told that the liquids are an acid, an alkali (base) and distilled water, but you are not told which one is which. Think about how you could use an indicator to identify these three liquids. Now copy the information below into your workbook and fill in the blanks with the correct word or words.

Litmus paper turns:

- ____ when dipped into an acid
- blue when dipped into a ____
- ____ when dipped into a neutral substance.

5 a Write simple instructions on how to make an indicator from red cabbage.

b Write brief notes on what you know about universal indicator.

6 Draw a simple diagram to explain the pH scale.

7 Classify these substances according to their pH values:

hydrochloric acid, vinegar, toothpaste, fizzy drink, lime juice, tap water, milk, toilet bowl cleaner, curry powder, bleach, dishwashing liquid, black tea.

- Record your answers in a table like the one below.
- Classify each substance as strong acid, weak acid, neutral (or nearly so), weak alkali or strong alkali.

Substance	Colour of universal indicator	pH value	Classification
lime juice	pale orange	4	weakly acidic

8 Read the paragraph below. Copy it into your workbook and fill in the blanks with the correct word.

In the middle of the pH scale, with a pH value of 7, are ____ substances.

- An acid is a chemical that will ____ a base such as an alkali.
- A ____ is a chemical substance that will neutralise an acid.
- ____ are example of bases. An alkali is the chemical opposite of an acid; it is a chemical that when mixed with an acid can ____ , or cancel out, the acid's effect.

9 Out of all the metals you have tested, which metal reacts most strongly with water? Which metal reacts least strongly?

10 Write down two factors (things) that affect the solubility of a substance.

11 Read the paragraph below. Give your own explanation for what happened.

20 g of sodium chloride was added to 20 ml of water while being heated and stirred. The water was heated continuously while a further 60 g of sodium chloride was added until it dissolved. The mixture was allowed to cool. Crystalline structures formed in the mixture and settled on the bottom of the beaker.

12 Write the word equations and chemical equations for the following reactions:

a hydrochloric acid reacting with sodium hydroxide

b sulphuric acid reacting with magnesium

c hydrochloric acid reacting with calcium carbonate.

Glossary

abiotic – non-living; abiotic factors include light, water and temperature
absorb – to take up; for example, a sponge absorbs water
accelerate – to speed up
acceleration – a change or increase in speed over time
acid rain – water droplets or rain containing high concentrations of sulphuric acid and nitric acid; acid rain is formed by sulphur dioxide and nitrogen dioxide dissolving in water vapour in the air; it damages soil and plant life, as well as buildings and statues
acid/acidic – a solution that has a pH of less than 7. An acid will neutralise a base such as an alkali.
action force – a force acting on an object; for example, a boy pushing against a wall
aerodynamic – designed to reduce air resistance, such as the skin-tight suit worn by cyclists
air – the mixture of gases that we breathe; air makes up the atmosphere and is a mix of nitrogen, oxygen, carbon dioxide and the noble gases
air pollution – the release of chemicals or substances into the air, which disturbs the balance in the atmosphere and can cause harm to the environment
air quality – a measure or description of how clean or dirty the air is
air resistance – the force of air pushing against a moving object
alkali/alkaline – a solution that has a pH of more than 7. An alkali will neutralise an acid and is the chemical opposite of an acid.
allotropes – different forms of an element, such as ozone (O_3) and oxygen (O_2), which are both made of oxygen atoms only. Graphite and diamond, which are both made up of only carbon, are also allotropes.
aneroid barometer – an instrument that measures changes in air pressure. Unlike many barometers that work using a liquid, such as mercury, aneroid barometers are dry barometers.
angle of incidence – the angle at which light strikes the surface
angle of reflection – the angle at which light bounces off the surface
annular eclipse – a solar eclipse in which the Moon appears smaller than the Sun and so the Sun is seen behind the disk of the Moon as a bright ring
aperture – the hole or opening of a camera that controls the amount of light let in
apparent depth – water looks shallower than it really is when viewed from above, because of the bending or refraction of light in the water
aquaplaning – skimming or sliding when there is a thin layer of water on the road, which prevents the tyres of a car making contact with the road
aquarium – a glass tank that water plants and animals are kept in
area – the size of a surface. Area is calculated as width × length.
argon – a noble gas that is used in electric light-bulbs to stop the filaments from burning
asterism – a recently identified pattern of stars such as the Summer Triangle
asteroids – small planet-like objects and rock-like bodies, which could be the remains of a shattered planet
astronaut – a person who is trained to pilot a spacecraft or serve as a crew member aboard a spacecraft
astronomy – the science of studying the stars and the planets
atmosphere – the layer of gases surrounding and protecting Earth
atmospheric pressure – pressure caused by the weight of air around Earth
atom – the smallest particle of any chemical element that has all the properties of the element. All matter is made up of atoms.
atomic number – the number of protons in an atom. This number is different for each different element's atoms.

atomic weight – the average mass of an atom. It is also known as *atomic mass*, and is different for different kinds of atoms.

attraction – pulled towards. Gravity, which pull objects towards Earth, is an example of a force of attraction.

autotrophs – self-feeding organisms such as plants

bacteria – a group of single-celled organisms that are too small to be seen by the naked eye. Bacteria play an important role breaking down decaying matter.

barometer – an instrument used to measure changes in air pressure

base/basic – a chemical compound that will neutralise an acid, producing a salt and water only. Alkalis are examples of bases.

beam – a stream of light

bioaccumulation – the build-up of poisons in the food chain, becoming more concentrated higher up the food chain

biodegradable – can be broken down by living organisms and recycled back into the environment

biological control – a method of controlling pest populations by means of their natural predators rather than use of pesticides

bioremediation – the use of micro-organisms to degrade toxic substances into less harmful ones

biotic – living

black holes – collapsed stars whose gravitational field is so strong that not even light can escape them

camera – a device used to take photographs or pictures

capillary action – the ability of substances, such as string or porous brick, to draw up a substance, such as water, because of forces of attraction between the two substances

carbohydrates – a group of foods such as sugars and starch that are made up of carbon, hydrogen and oxygen. Plants make carbohydrates (glucose) from carbon dioxide and water by the process of photosynthesis.

carbon – a non-metal element that is found in living organisms and in fossil fuels such as crude oil

carbonates – compounds containing the carbonate ion CO_3^{2-}.

carbon cycle – the cycle in which carbon is released into the atmosphere as carbon dioxide and removed from the atmosphere to make organic molecules such as carbohydrates

carbon dioxide – the gas in the air that is used by plants for photosynthesis and is produced by cellular respiration

carnivore – an animal that eats meat, feeding on other animals

carrying capacity – the maximum number of individuals in a population that the habitat will support

catalyst – a substance or an enzyme that speeds up a chemical reaction without itself being used up or changed in the process

catalytic converter – a device fitted to a car exhaust to reduce toxic emissions. A catalytic converter breaks down toxic gases such as nitrogen oxide and carbon monoxide into harmless products.

cellular respiration – the process by which animals and plants combine glucose with oxygen to release energy

centrifugal force – the outward force experienced by an object moving in a circle.

CFCs – chlorofluorocarbons; a group of chemicals that contain chlorine and are known to damage the ozone layer. CFCs were used as propellants in aerosol cans and as coolants in fridges.

chemical bonds – the forces that hold atoms together

chemical reactions – see reactions

chlorophyll – the green pigment in chloroplasts that gives plants their green colour and captures the Sun's energy for photosynthesis

chloroplasts – the tiny structures in the plant cell where photosynthesis takes place

collision – when two objects or vehicles crash into each other

colony – a place of human settlement, far from home

colour spectrum – the range of colours of light; it includes the seven colours red, orange, yellow, green, blue, indigo and violet, and all the colours in between

combustion – burning. In the process, oxygen combines with fuel to release energy.

comet – ball of ice, gas and dust which has an unusual elliptical orbit around the Sun

commensalism – the relationship between two different organisms in which the one benefits and the other is not affected in any way

community – a group of different kinds of plants and animals that live together and depend on each other

competition – the struggle between animals and plants of the same group or different groups for food, space and light

compound – a substance made from atoms of different elements joined together by chemical bonds. Most matter is made up of a combination of different kinds of atoms.

compress – to squash or fit into a smaller space

concave – curving inwards, like the inside of a teaspoon

condensation – the change from gas to liquid. Water vapour in the air condenses into liquid droplets as it cools.

cones – the cells of the retina that are sensitive to colour

constellation – a pattern of stars, usually one that was identified by astronomers a long time ago, such as Orion

consumers – organisms such as animals that cannot make their own food

contact force – a force, such as friction, which involves objects that are in contact or touching

converge – the bending of light rays so that they come together in a point. A convex lens is a converging lens.

convex – curving outwards, like the back of a teaspoon

cornea – the transparent or see-through protective covering over the eye

corona – the circle of light around the Sun, which can be seen during a total solar eclipse; *corona* means *crown*

corrosion – when metals react with oxygen and/or other substances from the atmosphere to form compounds. When iron or steel corrodes, rust is formed.

corrosive substances – substances such as acids that dissolve or eat away other materials

crescent moon – the Moon in its sickle-shaped phase

crude oil – a fossil fuel that is refined to make petroleum

cryopreservation – the freezing of cells, tissues or living samples in liquid nitrogen to preserve them

decelerate – to slow down

decomposers – organisms such as fungi and bacteria, which break down complex food molecules into simple ones and recycle nutrients back into the soil

deforestation – the destruction of forests by removing trees to prepare the land for farming or building

denitrification – the process in which bacteria turn nitrates in the soil into nitrogen gas

detritivores – organisms that feed on the refuse and dead food in an ecosystem

dispersion – the splitting of white light into colours by refraction

diverge – the bending of light rays so that they spread out. A concave lens is a diverging lens.

DNA – deoxyribonucleic acid; a complex molecule in the nucleus of each living cell that carries genetic information called genes

Earth – our home planet, also known as the blue planet because there is more sea than land, and so Earth looks blue when viewed from space

ecology – the study of ecosystems or the environment

ecosystem – a living system made up of plants and animals and their environment

ecotourism – a form of tourism for people who care about and enjoy the natural environment

elastic potential energy – the energy stored in a stretched rubber band or a coiled spring

elasticity – stretchiness. An elastic object or material goes back to its original shape when the force that changes its shape is removed.

electromagnetic spectrum – a range of energy-carrying waves that includes radio waves and light waves

electron – the negatively charged particles of an atom. Electrons orbit the nucleus of an atom.

electron shells – the different energy levels that the electrons of an atom occupy

electrostatic force – a force of attraction or repulsion between charged objects

element – each type of atom in the periodic table of elements; also a pure chemical substance made up of only one type of atom

elliptical – the shape of a squashed circle

endangered species – those groups of plants or animals that are in danger of becoming extinct because their numbers are very low

endemic – found only in one place or area and nowhere else
endoscope – a flexible optical fibre tube that is used to examine parts inside a patient's body without having to operate
epiphytes – plants such as some mosses, orchids and bromeliads, which grow on other plants, such as trees, for physical support
eutrophication – nutrient pollution of land or water that causes plants or algae to grow and decompose very quickly, using up oxygen supplies
evaporation – the change from a liquid to a gas; for example, water evaporating into water vapour
evolution – the process by which new plants and animals develop from existing plants and animals by adapting to changes in the environment
expand – to spread out to fill a larger space. Gases expand to fill the container they are in.
extension – stretching
extinct – plant or animal species that have died out completely
eyepiece lens – the lens of an optical instrument, such as a telescope, that magnifies the image; usually just called the eyepiece
fertilisers – chemicals rich in nitrates and phosphates that are added to the soil to make plants grow well
fibre optics – the engineering and technology that makes use of optical fibres, which are very thin fibres that carry light
filter – a translucent material, such as Cellophane, which lets only some light pass though. A red filter lets only red light through.
focal length – the distance from the centre of a lens to the focal point; it is a measure of how strongly the lens focuses light
focal point – the single point at which all light rays passing through a converging lens meet
food chain – a group of living things that depend on each other for energy. A food chain is ordered according to who eats who – it begins with a plant, which is eaten by an animal, which is in turn eaten by another animal.
food web – a complex system of feeding relationships, made up of food chains
force – a push or a pull that can move an object, change its shape, and transfer energy. Gravity and friction are examples of forces.
friction – a force that resists the movement of an object and produces heat. Friction takes place between two surfaces that are in contact or rub together.
fungi – a group of organisms, such as mushrooms and mould, that reproduce by means of spores. The singular of fungi is fungus.
g – acceleration due to gravity, which is 9.81 m/s^2
galaxy – a system of stars held together by gravitational forces
gas – a substance or state of matter that spreads out to fill all the space available; many gases are colourless, for example air, which is a mixture of gases.
geographical information satellite (GIS) – a satellite system that collects and analyses information about Earth's surface and atmosphere. This information might be used for map-making or for looking at weather patterns.
geostationary orbit – an orbit above Earth's equator, which matches the speed at which Earth spins so that the satellite stays in the same position above Earth. Communication satellites follow geostationary orbits.
gibbous moon – the Moon in its phase between quarter moon and full moon
global dimming – the decrease in the amount of sunlight that reaches Earth because of the increased amount of water in the air due to evaporation
global positioning system (GPS) – small devices that receive radio signals from a network of satellites circling Earth and use these signals to find their position. These devices are portable or fitted into aircraft, boats and cars.
global warming – the small but steady increase in Earth's temperature due to the build-up of greenhouse gases, such as carbon dioxide, in the atmosphere
glucose – a simple sugar, which is made by green plants from carbon dioxide and water and is broken down during cellular respiration
gravitational force – see gravity
gravity – the invisible force with which Earth pulls objects towards it, and the force of attraction between all objects with mass. Gravity is the force that governs the universe and our solar system, giving objects weight and making the planets circle the Sun.
greenhouse effect – the trapping of Earth's heat by gases
greenhouse gases – gases such as carbon dioxide, methane and water vapour, which trap heat around Earth

habitat – the place where plants grow and animals live

helium – one of the noble gases and the second most common and second lightest element. This light-weight gas is often used to fill weather balloons and it is formed by nuclear fusion in the stars and the Sun.

herbivores – animals that feed only on plants

heterotrophs – organisms which cannot make their own food, and so need complex organic substances for food

Hooke's Law – this law states that the extension of a spring or the distance it stretches is directly proportional to the force acting on it. In other words, how much an elastic material stretches depends on the amount of force you apply.

Hubble telescope – officially called the Hubble Space Telescope; this giant telescope is named after the astronomer Edwin Hubble and it orbits Earth, sending back images of the solar system and distant galaxies

humidity – the amount of water vapour in the air. Humidity depends on temperature, because the warmer the air, the more water vapour it can hold.

hydraulic – a system that uses liquid pressure to transmit a force. Car brakes and some lifts operate by hydraulics.

hydrogen – the simplest and most common element, which makes up the most of the Sun. The hydrogen nucleus contains a single proton.

hygrometer – an instrument used to measure humidity

hygroscopic – a hygroscopic substance attracts or absorbs water vapour from the air

hyphae – the thread-like structures that grow from a fungus

image – the view or picture formed by a lens or mirror

indicators – substances whose colour depends on the pH of the solution they are in

indigenous – local or natural. Indigenous plants and animals belong naturally to the region and haven't been imported or introduced.

inert – unreactive. The noble gases are also called the inert gases because they are unreactive.

inertia – the tendency of an object at rest to stay at rest and of an object in motion to stay in motion

interaction – acting together; actions or relationships between animals and plants

International Space Station (ISS) – a large research space station that orbits Earth and is still in the process of being put together; many countries are involved in the project

ions – atoms that have gained or lost electrons and so have a negative or positive electric charge

iris – the coloured ring of the eye around the pupil

Jupiter – a giant gas planet with many moons and a swirling storm cloud known as the Great Red Spot

kaleidoscope – a tube-like instrument or toy that uses mirrors and small coloured pieces to make patterns from the reflections

kinetic energy – energy of movement. Moving objects have kinetic energy.

Kuiper belt – a broad region beyond Neptune that is made up of ice bodies

landfill – a dump site where waste is buried as a way of disposing or getting rid of rubbish or refuse

lens – a curved piece of glass or plastic designed to refract or bend the light. The lens of the eye is made of an elastic substance that can change its shape.

light ray – a thin line of light

light year – the distance that light travels in one year, which is about 9 500 000 000 000 or 9.5 trillion kilometres. This unit is used to measure the very large distances in space.

litmus – a powdery water-soluble dye, obtained from lichens, which turns red in acids and blue in bases

long-sighted – able to see far-away objects clearly, but not objects that are nearby

loss of habitat – when the living place of plants or animals is destroyed or when animals are forced to move from their living place because it is taken over by humans

low polar orbit – an orbit at right angles to the direction of the Earth's rotation, circling the North and South poles. Weather satellites follow low polar orbits so that they can cover the whole planet as Earth rotates.

luminous – shining or producing its own light; an example of a luminous object is a candle or a star

lunar eclipse – the event when Sun, Earth and Moon line up with Earth in the middle, blocking out the view of the Moon. A lunar eclipse can only take place during a full moon.
magnetic force – force of attraction or repulsion between moving electrically charged particles
magnifying glass – a lens that makes objects look bigger than they are, often used for reading small writing
marine ecosystem – a living system of plants and animals in seawater
Mars – a red-coloured terrestrial planet, which is in some ways similar to Earth
mass – the amount of matter in an object; it is measured in kilograms and is constant
Mercury – the small, hot terrestrial planet closest to the Sun
meteor – also known as a shooting star; a meteor is a meteoroid which burns up as it passes through the Earth's atmosphere
meteorite – a meteoroid which lands on Earth before it burns up completely
meteoroid – any asteroid that is smaller than 10 metres in diameter and contains specks of rock
microgravity – very weak gravity, which produces a state of weightlessness
micro-organisms – very small organisms such as bacteria, which can only be seen under a microscope
Milky Way – the spiral-shaped galaxy to which Earth and its solar system belong
mirage – a shimmering caused by the light bending as it passes though a layer of hot air just above the surface of a tar road, for example
mirror – a surface that gives a good reflection, producing a virtual image. A mirror is usually made of glass coated with a thin layer of aluminium or silver.
mitochondrion – the tiny structure in a cell where cellular respiration takes place; the plural of mitochondrion is mitochondria
molecule – a particle made up of two or more atoms joined together. A molecule of oxygen is made up of two oxygen atoms, and a molecule of carbon dioxide is made up of one atom of carbon and two atoms of oxygen.
momentum – a measurement of motion, which depends on the mass and the speed of the object
monoculture – the growing of one crop only on a large scale
moon – a planet's natural satellite. Earth has one moon and Jupiter has many.
motion – movement
mutualism – a relationship between two different organisms that benefits both of them
NASA – National Aeronautics and Space Administration, the agency for the United States' space research and exploration programme
neon – a noble gas that is used in neon advertising lights
Neptune – the furthest planet from the Sun
neutral solution – a solution that is neither acidic nor alkaline. It has a pH of 7.
neutralisation – when an acid reacts with an alkali or insoluble base to make a neutral solution
neutrons – the particles of an atom that have no charge
newton (N) – a unit of force measurement. On Earth, each kilogram has a weight of 9.8 newtons.
newton meter – an instrument used for measuring the size of a force, also called a spring balance
nitrogen – the most common gas in air; unlike oxygen, it is unreactive
nitrogen cycle – the cycle in which nitrogen is circulated or exchanged between the atmosphere, as nitrogen gas, and Earth and living organisms, as ammonia and nitrates
nitrogen dioxide – NO_2, the brown, poisonous gas which gives smog its dirty colour
nitrogen fixation – the process in which a special group of bacteria turn nitrogen gas from the atmosphere into ammonia
nitrogen monoxide – NO, the poisonous gas produced by car engines and power plants
noble gases – a group of unreactive gases which make up 1% of the air. The noble gases are helium, neon, argon, krypton, xenon and radon.
non-contact forces – forces which act at a distance and don't involve objects touching, such as gravitational force
non-luminous – an object that does not give off its own light, such as a planet
normal – a perpendicular line that we draw at right angles to a surface

North Star – the 'pole star', which was used for navigation by sailors long ago. Because the North Star is on the same axis as the Earth's rotation, its position in the night sky is fixed.

nuclear energy – the energy released by reactions that involve the nuclei of atoms

nuclear fission energy – the energy produced when the nucleus of an atom is split

nuclear fusion energy – the energy produced when hydrogen nuclei fuse together in the reactions that take place in the Sun and other stars

nucleus – the dense centre of an atom that contains protons and neutrons

objective lens – the larger lens of an optical instrument, such as a telescope, that focuses the image of a far-away object

omnivores – animals that eat both plants and animals

opaque – not clear or see-through, because light is blocked. Examples of opaque materials are metal and cardboard.

optical fibres – very thin fibres of glass that carry light by total internal reflection

optical illusion – a trick of the light, making us 'see' something that isn't really true

optical instruments – instruments, such as microscopes, binoculars and telescopes that are used for looking at objects

orbit – the circular path that a planet follows around the Sun, or the path that an object, such as the Moon or a satellite, makes around another object, such as Earth. The time it takes Earth to complete an orbit around the Sun is a year.

organic farming – natural farming methods that don't use pesticides and artificial fertilisers

oxide – a substance or compound that contains oxygen. Oxides form when other elements combine with oxygen in the air.

oxygen – the gas in the air that is produced by photosynthesis and is needed for cellular respiration; it is the second most abundant gas in the mix of gases that make up the air

ozone – O_3, a gas with molecules made up of three atoms of oxygen; it is formed by ultraviolet light reacting with oxygen in the air and it makes up a thin layer in the upper atmosphere that protects Earth from the Sun's harmful UV radiation; at ground level, ozone is a pollutant

parasitism – the relationship in which an animal or plant (the parasite) lives on another plant or animal (the host); only the parasite benefits, and the host is often harmed

particulate pollution – usually solid particles such as dust, soot or pollen, which are small enough to be carried in the air and can cause harm when they are breathed in

pascal (Pa) – the unit of force, equal to N/m^2

penumbra – the lighter, fuzzy region around a shadow where the light is not blocked completely

periodic table – a table of the elements, arranged in order of their atomic number

periscope – a device made with two mirrors or prisms positioned at an angle, making it possible to see around corners or above the surface of the sea from a submerged position

pesticides – chemical poisons used to kill insect pests

pH scale – a scale that tells you how acidic or alkaline a solution is

photosynthesis – the process by which plants make their own carbohydrate foods (glucose) from carbon dioxide and water, using energy from the Sun. Oxygen is a product of photosynthesis.

pigment – a coloured substance; chlorophyll is a green pigment, which means that it reflects green light and absorb all other colours of light

planet – a large body that orbits a star. A dwarf planet, such as Pluto, is a planet which is not big enough to clear the neighbourhood around its orbit of small bodies and debris.

pollutant – a chemical substance in the air, water or soil that can cause harm to the environment

population – a group of individuals of the same kind or species that all live in the same place

precipitation – the change of water vapour in the atmosphere into a liquid form such as rain; this is a type of condensation

predation – the relationship between two kinds of animal, in which one hunts, kills and eats the other

pressure – force acting over area; a measure of how concentrated a force is

primary colour – one of the three main colours of light, which are blue, green and red

primary consumers – animals in the food chain that eat grass and other plants

prism – a triangular-shaped piece of glass that splits white light into its colours

producers – organisms, such as plants, that make their own food

products – the new substances that are produced in chemical reactions

proteins – a group of foods that contain nitrogen. Hormones and enzymes are proteins.

proton – a positively charged particle in the nucleus or centre of an atom

pupil – the opening in the eye that controls the amount of light let in

pyramid of numbers – a triangle-shaped diagram showing the different feeding levels in a food chain, with the numbers of plants or animals in each level

rainbow – an arc of colours in the sky, caused by the splitting of white light into all its colours by raindrops, which act as little prisms

Rayleigh scattering – the scattering of blue light by the molecules of oxygen and nitrogen in the air, making the sky look blue

reactants – the substance(s) that you begin with in a chemical reaction

reaction (or chemical reaction) – a process in which substances have an effect on each other and new substances are produced

reaction force – a force in response to an action force; for example, when a boy pushes against the wall, the wall pushes back with equal force

real image – an image which can be seen on a screen

recycle – to reprocess materials such as cardboard, paper, glass and tin into new products

reflect – to bounce back from the surface

reflection – the bouncing back of light rays in the same direction that they came from; the image seen in very smooth surfaces, such as a mirror or still water, when the light is reflected

refraction – the bending of light when it moves from one medium to another

repulsion – pushed away. The like poles of two magnets pushing against each other is an example of a force of repulsion.

respiration – see cellular respiration

retina – the back of the eye, where the image falls

rhizobia – the group of nitrogen-fixing bacteria that live on the roots of legume plants and turn nitrogen in the air into ammonia

RNA – ribonucleic acid; similar to DNA, but it acts as the message between the DNA code of a gene and its protein product

rods – the cells of the retina that are sensitive to the amount of light

rotate – to spin around a central axis. Earth rotates around its axis every 24 hours.

salt – a compound produced when you neutralise an acid with an alkali or an insoluble base. Copper sulphate and calcium carbonate are examples of salts.

satellite – an object that orbits a planet. The Moon is a natural satellite, and communication satellites and weather satellites are artificial satellites.

Saturn – a giant gas planet with bright rings, which are made up of lumps of ice

scavengers – animals, such as vultures and hyenas, that feed on the carcasses or remains of dead animals

secondary colour – a colour made by mixing two primary colours together. For example, yellow is a secondary colour made by mixing red and green light.

secondary consumers – animals in the food chain that eat other animals

shadow – a shape or 'picture' made where light is blocked by an opaque object

short-sighted – able to see nearby objects clearly, but not objects that are far away

slash-and-burn agriculture – a method of clearing forests by cutting and burning to make open fields for farming

sliding friction – friction between a moving object and the surface it is in contact with

small solar system bodies (SSSBs) – a group of bodies in space that includes asteroids, Kuiper belt objects and comets

sodium chloride – the chemical name for common table salt

soil erosion – the loss of soil, blown away by wind, or washed away by water. Soil erosion takes place when plants are damaged or destroyed and they no longer hold the soil in place.

solar eclipse – the event when the Moon passes directly between the Sun and the Earth, blocking out the view of the Sun

solar energy – also known as solar power, it is energy produced by the Sun, which can be used for human purposes such as heating and producing electricity

solar flare – a violent explosion in the Sun's atmosphere, which is caused by the Sun's strong magnetic field

solar system – the Sun, and all the objects which circle it. The solar system is made up of the eight planets, their moons, the dwarf planets, asteroids and comets.

solubility – how much of a solute can dissolve in a solvent at a particular temperature

solution – one substance (often a liquid, such as water) with another substance dissolved in it

solvent – substances that dissolve things, or a substance in which other substances are dissolved, often a liquid

solute – the dissolved substance in a solution

spores – tiny seed-like reproductive bodies produced by fungi and simple plants like ferns

spring balance – an instrument used for weighing an object or measuring the size of a force

star – a giant light-emitting body in space

static friction – friction between a stationary object and the surface it is resting on

stomata – the small holes or 'breathing' pores on the surface of a leaf, through which water is lost and carbon dioxide and oxygen enter and leave

stratosphere – the second layer of the atmosphere, about 50 km above Earth

sulphur dioxide – SO_2, a colourless gas which is produced by the burning of fuels, such as coal, that contain sulphur

supernova – an exploding star

symbiosis – a close relationship between two different kinds of plant or animal

technology – the use or application of science to solve problems or make products to meet the needs of society

telescope – an instrument used to look at far-away objects such as the planets and the stars. A telescope is designed to gather and focus the light from far-away objects so that we can see them more easily.

terrarium – a closed glass container that plants are grown in

tides – the changes in the level of the sea caused by the gravitational pull of the Moon on the Earth. The time between high tide and low tide is about six hours.

total internal reflection – when light is bent or refracted in such a way that it is reflected back

traditional medicine – the treatment of illnesses using natural methods and natural medicines that are often made from plants

transect survey – using a sample strip of ground to study plant and animal populations

translucent – partly see-through, because only some light is let through. Examples of translucent materials are tissue paper, Cellophane and stained glass.

transparent – see-through, because light is let through. Examples of transparent materials are glass and Perspex.

transpiration – the loss of water from the leaves of plants

troposphere – the 16-km-thick layer of air that is closest to Earth

ultraviolet (UV) light – it is part of the electromagnetic spectrum and has a shorter wavelength than visible light. The Sun emits visible light and UV light.

umbra – the dark centre part of a shadow where all light is blocked

universal indicator – an indicator that changes to many different colours depending on the pH of the solution that it is in

universe – all of space

Uranus – a giant blue gas planet

valency – the number of electrons in the outermost shell of an atom; it is a measure of how many chemical bonds the atom can form

Venus – a terrestrial planet and Earth's nearest neighbour

virtual image – an upright image from which the light rays seem to come. A mirror image is an example of a virtual image.

waning – the lit face of the Moon gets smaller, moving towards new moon

water – a molecular compound of hydrogen and oxygen

water cycle – the endless movement and cycling of water in different states between the land and the atmosphere

water vapour – water in a gas state or form

wavelength – the distance between one wave crest or peak and the next. The shorter the wavelength, the more energy the wave carries; blue light has a shorter wavelength than red light.

waxing – the lit face of the Moon gets bigger, moving towards full moon

weight – the gravitational pull of Earth on objects; weight is measured in newtons and changes if the force of gravity changes

weightlessness – the feeling of having no weight because of reduced gravity. Astronauts in space experience weightlessness.

wetland – a marshy area of land covered in water

Index

T

U

V

W

Y

Acknowledgements

The authors and publisher would like to thank the following copyright holders for permission to reproduce their photographs:
SPL = Science Photo Library; t = top; b = bottom; l = left; r = right.

©birdseen.co.uk/Ian Hillery 8r; ©Bob Turner/Alamy 30r; ©Buzz Pictures/Alamy 157l; ©Carol Ramjohn 22r; ©Helen Collett 157; ©iStockphoto.co./Andrey Volodin 96tl; ©iStockphoto.co./Joseph 113t; ©iStockphoto.com/Anthony Collins 142t; ©iStockphoto.com/Dawn Hudson 129b; ©iStockphoto.com/Pattie Calfy 104t; ©iStockphoto/Duncan Walker 125b; ©mike lane/Alamy 153t; Alexis Rosenfeld/SPL 15l; Alfred Pasieka/SPL 113m; Andrew Lambert Photography/SPL 57mb, 136t, 147t; Bildagentur-online/Th_foto/SPL 35b; Bob Gibbons/SPL 109b; Bruce Coleman Inc./Alamy 15r; BSIP, Delacourt/SPL 38l; BSIP, Keene/SPL 140b; Carl Schmidt-Luchs/SPL 116l; Charles D. Winters/SPL 44bl; Chris Butler/SPL 87m; Claude Nuridsany & Marie Perennou/SPL 140m; Coneyl Jay/SPL 62; Cordelia Molloy/SPL 47t; David Aubrey/SPL 16tr; David N. Davis/SPL 63r, 105tl; David Parker/SPL 126m; DavidSimonsB-6940Septon(daseducatief@yahoo.com 23, 27t, 32, 34t, 35t, 38r, 43m, 44tl, 65t, 60t, 55m, 69b, 71t, 72, 85tr,97, 108b, 109r, 112b, 117tr, 128m; Diccon Alexander/SPL 23r; forty40 photography/Alamy 167; Dr Fred Espenak/SPL 94r; Dr George Gornacz/SPL 7br; Dr Jean Lorre/SPL 100br; Dr Jeremy Burgess/SPL 45b; Dr M.A. Ansary/SPL 66tr; Dr Morley Read/SPL 7tr, 8tr, 8m, 9, 19r, 30l; Dr P Marazzi/SPL 50m, 10l; Dr Tim Evans/SPL 49; Dr. John Brackenbury/SPL 129m; E.R. Degginger/SPL 18bl, 26b; Edward Kinsman/SPL 148t; Eric Erbe/SPL 16bl; Erich Schrempp/SPL 114m; European Space Agency/SPL 88r; Eye of Science/SPL 16tl; Fletcher & Baylis/SPL 16br; Gary Hincks/SPL 92t; Gary Parker/SPL 63l; Geoff Tompkinson/SPL 51; Gustoimages/SPL149b; H.E. Bond/E. Nelan/M. Barstow/M. Burleigh/J.B. Holberg/NASA/ESA/STSci/SPL 83t; Hugh Spencer/SPL 53; Jany Sauvanet/SPL 26t; Jeremy Bishop/SPL 157r; Jerry Lodriguss/SPL 95b; Jim Edds/SPL 33; Jim Reed/SPL 43b; Jim Zipp/SPL 7bl, 18tr; John Chumack/SPL 83b; John Mead/SPL 85br; John Reader/SPL 36b; John Sanford/SPL 78, 79l, 92b; Ken Cavanagh/SPL 138b; Kent Wood/SPL 114t; Mark Garlick/SPL 87t, b; Marlin E. Rice/Agstock/SPL 31t; Martin Bond/SPL 127b; Martin Dohrn/SPL 14; Mehau Kulyk/SPL 44m; NASA/ESA/STSCI/E. Karkoschka, U. Arizona/SPL 80m, 88m; NASA/JPL/Cornell/SPL 100bl; NASA/SPL 66b, 69tl, tr, 80lr, 81, 88l, 93m, 96bl, 98r, 99, 100, 101t; Paul Avis/SPL 110t; Paul Rapson/SPL 57t; Paul Silverman/Fundamental Photos/SPL 47b; Peter Chadwick/SPL 18m; Philippe Psaila/SPL 37b, 108t; Photo Researchers/SPL 27b; Prof. David Hall/SPL 17r; Rafael Macia/SPL 139; Ralph Eagle/SPL 128t; Rev. Ronald Royer/SPL 94l; Ria Novosti/SPL 98l; Rod Planck/SPL 7tl; Royal Astronomical Society/SPL 87m; Sam Ogden/SPL 155b; Science Photo Library 82t, b; Scimat/SPL 31b; Scott Bauer/US Department of Agriculture/SPL 50r; Sheila Terry/SPL 71b; Simon Fraser/SPL 37t; Stephen J. Krasemann/SPL 10r; Ted Kinsman/SPL 104m; The Bigger Picture/Alamy 150b, 156t; Tony Craddock/SPL 8m; Will & Deni McIntyre/SPL 117tl; iStock 164, 165, 171, 172; Wendy Lee 162, 169; Scott Camazine/SPL 163.

Every effort has been made to trace copyright holders. Should any infringements have occurred, please inform the publisher who will correct these in the event of a reprint.